ASPECTS OF ZOOGEOGRAP

ASPECTS OF ZOOGEOGRAPHY

by

Paul Müller

Dr. W. Junk b.v., Publishers, The Hague, 1974

ISBN 90 6193 023 5

Dr. W. Junk b.v., Publishers, The Hague
Cover design: Max Velthuijs
Printed in the Netherlands
by Koninklijke Drukkerij Van de Garde B.V., Zaltbommel

CONTENTS

PREFACE

Zoogeography aims to explain the structure, function and history of the geographical ranges of animals. The absence or presence of a species in a given place has ecological as well as historical causes. It is therefore a mistake to suppose that reconstructing the phylogenetic connections of a taxon will by itself give a definite picture of how its range originated. A purely ecological interpretation of the range could be equally misleading if it did not take into account the population-genetic structure underlying the geographical range. Phylogenetic systematics, population genetics, autecology and synecology have all their own methods, none of which can be substituted for another, without which a range cannot be studied or interpreted.

The present book covers only certain aspects of the wide field of zoogeography. These are in the form in which they were crystallised in the course of innumerable discussions with my teachers, my colleagues at home and abroad and my fellow workers, postgraduates and students at Saarbrücken, as well as in the zoogeographical part of may basic lectures on biogeography for the year 1973–1974. The chief emphasis is laid on the genetic and ecological macrostructure of the biosphere as an arena for range structures and range dynamics, on urban ecosystems, which have hitherto been grossly neglected, and on the most recent history of ranges (the dispersal centre concept). The marine and fresh-water biocycles, on the other hand, have been dealt only briefly. I have attached great value to references to the most recent literature which will give further insights into the subject, although I cannot claim that all important new works have been cited.

'Aspects of Zoogeography' can only show the path that we have followed up to the present and indicate the destination that we seek to reach. This path was mapped out in discussions with my teachers G. DE LATTIN, W. F. REINIG and J. SCHMITHÜSEN as well as with my colleagues and students in Saarbrücken. It was made negotiable by my wife who, by taking over and dealing with innumerable day-to-day problems, provided me with the necessary elbow-room so that I could devote myself to my hobby. To all these I express my profound gratitude. I should also like to thank Mrs. A. KONZMANN for typing the manuscript and Mr. CH. BENZMÜLLER and Mr. W. PAULUS for preparing the range maps.

Saarbrücken, August 1974. P. MÜLLER

ZOOGEOGRAPHY AND BIOGEOGRAPHY

Zoogeography is the subdivision of biogeography that studies the faunal make-up of landscapes and regions, the evolution and present-day dynamics of the geographical ranges of animals and the mutual relations of these ranges with mankind (MÜLLER 1972). The occurrence of an animal or of a plant in a particular place on earth will have population-genetic, ecological and historical causes. Organisms can only occur in a place if their ecological valency – i.e. the complex of living conditions in which a species or a single animal can survive (HESSE 1924) – does not conflict with the total environment of that place. For this reason animals and plants are living indicators of environmental characteristics. By means of their regions of distribution they define areas where environmental conditions are uniform or similar.

Biologists expect that zoogeography will contribute methodologically to the theory of evolution (DE LATTIN 1967). Nonetheless the geographical aim of zoogeography is to throw light on spatial characteristics and spatial effects of landscape. As opposed to geobiology, the ultimate concern of biogeography is not the organisms or the living communities themselves, but the landscape in which these organisms live (SCHMITHÜSEN 1957, MERTENS 1961, MÜLLER 1972, 1973, WALTER 1973). The geographer does not become a biologist when he applies special zoogeographical methods, any more than a biologist becomes a chemist or physicist when he applies chemical or physical mehods. The sciences are separated from each other, not by the methods they apply, but by what they aim to find out.

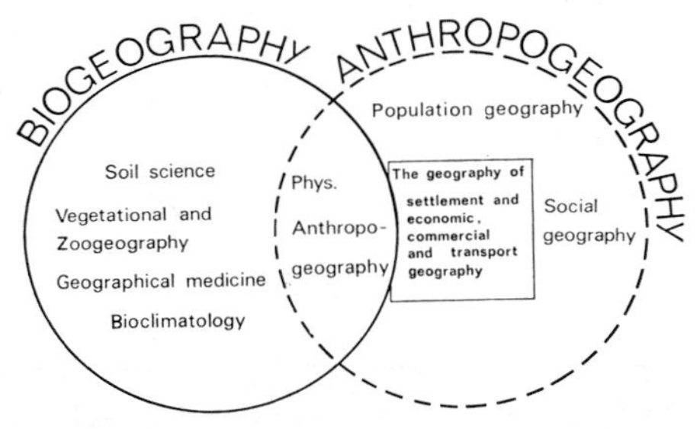

Fig. 1. The position of zoogeography relative to human geography, within biogeography.

The valency of a particular species in its environment is an ecological problem. But knowledge of this ecological valency is a necessary condition for the causal interpretation of a geographical range, whose study is the concern of biogeography (MÜLLER 1972).

A taxon is a group of organisms which can be recognised as a coherent unit at any particular level of an hierarchical classification; thus the common viper is a taxon, as also are the Viperidae, the Serpentes, the Reptilia, the Vertebrata and the Chordata. Now some taxa are strictly bound to a specific environment. Since they are dependent on the interaction of the factors that affect the particular part of the earth where they live, such taxa can be used as indicators for the external life conditions that operate at that place and which act as limiting factors for regions with the same or similar environmental conditions (MÜLLER 1972, SMARDA 1970). Thus the lichen *Parmelia physodes*, which can be widespread in our climate, is highly sensitive to various emissions, such as SO_2. In concentrations of 0.11 mg SO_2/m^3 of air this lichen generally disappears (SCHÖNBECK 1969, STEUBING 1970, KIRSCHBAUM 1973, KUNZE 1972). Its pattern of distribution therefore produces a mosaic which can be used in mapping the climatic effects of towns. Many other plant species are very useful as indicators of particular minerals in the search for new ore bodies (BROOKS 1972). Thus typical serpentine, zinc and copper plants are known (LYON et al 1970, ERNST 1967). Similar cases exist among animals. The use of saprobic species to evaluate the quality of water is based on the ecological valency and spatial distribution of bio-indicators (KOLKWITZ 1950). Usefulness as an indicator depends on a narrow ecological valency, such as characterizes stenotopic species strictly bound to a particular biotope. Eurotopic species, on the other hand, can exist under a great variety of conditions (cf. e.g. FREEMAN 1972, THOMAS 1972).

Within one species there can be a large or a small intraspecific variability in ecological valency. Populations that live near the limits of geographical range

Species	Typical Fodder plants	Fodder plants in Northwestern Germany
Orygia ericae	Salix, Myrica, Coniferen	Calluna
Dasychira fascelina	Sarothamnus, Salix	"
Malacosoma castrensis	Euphorbia, Potentilla	"
Lasiocampa quercus	Sarothamnus, Rubus, Prunus	" most important
Eudia pavonia	Rubus, Rosa, Prunus u. a.	"
Rhyparia purpurata	Sarothamnus, Rubus u. a.	"
Rhagades pruni	Prunus	"
Gastropacha quercifolia	Prunus	Salix

Fig. 2. Specialisation of feeding habits at the limits of geographical range. An example of change in ecological valency.

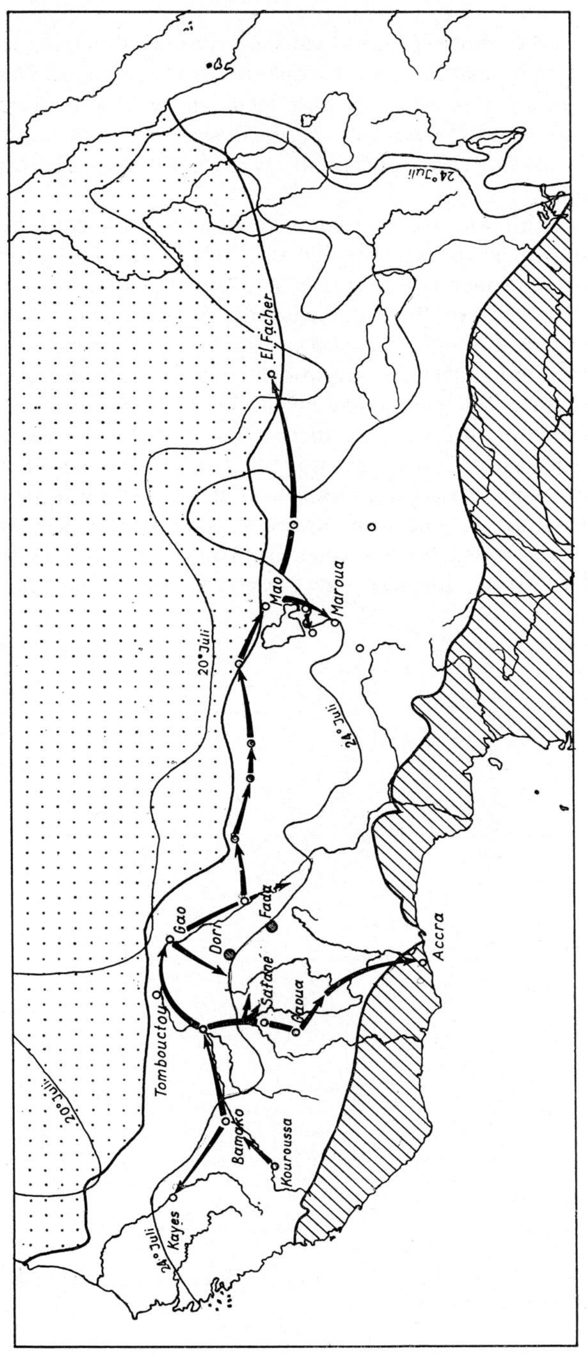

Fig. 3. Paths of dispersal of relapsing fever in African savanna biomes, controlled by the ecological valency of the vector (from MÜLLER 1972).

commonly have a different ecological valency to organisms from a central part of the range and males can have a different valency to females. Thus the purple hairstreak butterfly *Lasiocampa quercus* feeds in central Europe mainly on *Sarothamnus*, *Rubus* and *Prunus*, but the north-west German populations have specialized in *Calluna* as their food plant (WARNECKE 1936, HEYDEMANN 1943, DE LATTIN 1967).

By their distribution and spatial dynamics, pests and the causal organisms of diseases have influenced or even decided the fates of human cultures (MÜLLER 1974). One zoogeographer may concern himself with Chironomids and their significance for transoceanic land connections. Another may study the spatial dynamics of *Glossina*, *Anopheles*, *Aedes*, *Panstrongylus megistus* or *Triatoma infestans*. But the importance for geographical medicine of the animals that they choose to investigate will be completely different in the two cases. For the Chironomids seem to play no part in the distribution of disease, while the endemic occurrence of the other species is the cause of the regional distribution of the specific diseases of sleeping sickness, malaria, yellow fever and Chagas' disease. Compare in this connection: BARNETT 1960, BANTA & FONAROFF 1969, AUDY 1956, ASPÖCK 1965, BENTON 1959, BRUIJNING 1956, CHAMBERLAIN 1956, DAVIES 1965, FRITZSCHE, LEHMANN & PROESELER 1972, GIGLIOLI 1965, HESS

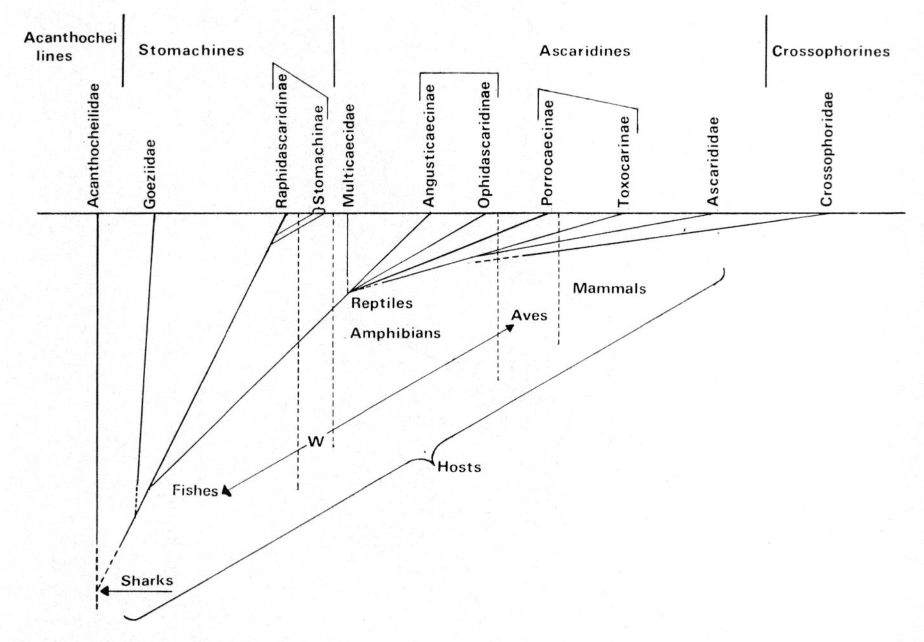

Fig. 4. The connection between the evolution of parasites (Nematodes) and the phylogeny of the host animals (after OSCHE 1958).

1956, Hopla 1961, Huba 1965, Hunter & Young 1971, Hurka 1967, Jackson 1965, Jahn 1965, Jusatz 1966, Klockenhoff 1969, Knight 1971, Knipling 1956, Lewis 1965, MacDonald 1965, Martini 1965, Müller 1972, 1974, Nash 1969, Nuorteva 1963, Pretzmann 1965, Reid 1965, Rodenwaldt 1925, Scherf 1969, Schmid 1969, Smetana 1965.

Geographical medicine represents a particular tendency in epidemiological research. Taking into account the results of microbiology, parasitology and immunology, and even the effects of atmospheric phenomena, it studies the distribution of diseases and epidemics in space and time on the whole earth (Jusatz 1964, 1972, Banta & Fonaroff 1969, Dever 1972).

Knowledge of the evolution of the species which belong to various regions of distribution throws light on the origin and history of our landscapes and continents, as well as on the course of world-wide climatic and vegetational change. But prerequisites for work of this sort are a knowledge of ecological valency, an assumption of its constancy and a consistent application of the principles of phylogenetic systematics in the sense of Hennig (1966) and Brundin (1966, 1972).

The Features of landscapes and of regions and the evolution of the animal that life in them affect each other mutually and illuminate each other mutually. The results of zoogeography contribute essentially to a deeper understanding of the evolutionary history of organisms, of our present-day landscape relationships, and of our own environment.

The Subdivisions of Zoogeography

Ecological and historical zoogeography are coherent units. This is the more true, according to De Lattin (1967), because what we today call history to a large extent represents the ecology of earlier times. Experimental and applied zoogeography, which are of very recent origin, have obscured, at least by the methods that they use, the distinction between biogeography and ecology. One can only concur with the sentiments of MacArthur & Wilson (1971) when they write: 'We both call ourselves biogeographers and recognize no real difference between biogeography and ecology' (see also Niethammer 1958, Niethammer & Kramer 1966, Økland 1956).

The field of research of zoogeography can be divided into a descriptive part, an causal part, and an applied part.

Descriptive zoogeography attempts to describe and arrange in order the confusing diversity of living phenomena in space. It uses for this purpose several sub-disciplines, i.e. chorology, faunistics, systematic zoogeography and biocoenotic zoogeography. Chorology seeks to understand as exactly and fully as possible the geographical ranges of organisms (cf. De Lattin 1967, Udvardy 1969, Adams 1970, Kaiser, Lefkovitch & Howden

1972). Faunistics attempts to write an inventory of species on the earth. Systematic zoogeography deals with the spatial distribution of the larger groups of animals such as birds or carnivores. Biocoenotic zoogeography on the other hand studies the distribution and dynamics of life-communities.

By the word biocoenosis (MÖBIUS 1877) is understood a life-community which constitutes the living part of an ecosystem (WOLTERECK 1928, TANSLEY 1935, ELLENBERG 1973).

A life-community and the place, or biotope, where it exists together form a mutually dependent whole which has its own dynamics. Individual animals will replace each other in a biocoenosis but, if in balance with its environment, the population system remains the same in its characteristic species. This condition of balance is known as biocoenotic equilibrium. Elucidation of the structure, changes, stability and distribution of particular biocoenoses is an essential task of ecological zoogeography.

Causal zoogeography can be divided into ecological, historical and experimental zoogeography. Ecological zoogeography investigates the ecological connection of the animal with the place where it lives. It thus attempts to understand the environmental factors which determine the area of distribution of a species. Historical zoogeography tries to explain the present-day areas of distribution of animals in the light of what is known of the origin and evolution of organisms and landscapes. Ecological and historical zoogeography are nevertheless not opposed to each other. Any argument about which of the two is more important is, as DE LATTIN said (1967), quite beside the point. For in actual fact both approaches are continually necessary for a full understanding of the causes of the distribution of organisms.

Experimental zoogeography develops experimental approaches to throw light on particular facts of distribution. The elucidation of the possibilities of passive dispersal of organisms (cf. SIMBERLOFF and WILSON 1969) and of their ecological valency (cf. STROHL 1921, THIELE and LEHMANN 1967, THIELE 1968) are important tasks for experimental zoogeography.

The most neglected part of zoogeography is **applied zoogeography**. In it the results of causal zoogeography are made useful to man. Its importance is immediately obvious where the subjects of research are parasitic, disease-carrying or economically important animal species. The history of domestication of our domestic animals presents a number of zoogeographical aspects (cf. e.g. UCKO & DIMBLEBY 1969).

Applied zoogeography also plays a decisive rôle in the biological control of pests. Recognition of the ecological valency and present-day distribution of the mosquito-fish *Gambusia affinis* was a basic necessity for its introduction into areas outside its natural range (MYERS 1965, FOWLER 1964, BIRIUKOV 1944, BOGDANOVICH 1935, CHACKO 1948). Knowledge of the relevant zoogeography is also important in the case of Diptera that serve as disease vectors (LAVEN 1959, 1969, 1972). The saprobic indicator system for evaluating the quality of water depends on the spatial distribution of particular indicator organisms and nowadays has important applications in regional planning (MÜLLER 1972). Future town planning, to be successful, will have to pay more attention to the spatial distribution of organisms in the town ecosystem (MÜLLER 1972, 1973). The subdisciplines of zoogeography listed above can be applied to all three of the great life environments of the biosphere – land, fresh-water and ocean. There is a zoogeography of the sea, of the inland waters and of the land.

Zoogeography is concerned only with the biosphere (LAMARK 1744–1828) as far as our present knowledge extends about the occurrence of life outside the earth. The biosphere is the living part of the geosphere. According to CAROL (1963) the geosphere includes the following: 1. Lithosphere; 2. Hydrosphere; 3. Pedosphere; 4. Biosphere; 5. Atmosphere; 6. Anthroposphere.

The biosphere envelops the globe in the sea, the land and the air as a layer not more than 20 km in thickness (SUESS 1909, VERNADSKIJ 1967). It is a heterogeneous system, which together with the inorganic part of the geosphere, constitutes the highest ranking ecosystem. The crust of the earth is the birthplace of the biosphere (SCHMITHÜSEN 1968, STUGREN 1972), and in this aspect can be referred to as the biogenosphere (ZABELIN 1959) or probiosphere (KOVALSKIJ 1963). The living part of the earth's crust constitutes only 0.1%, or between 10^{14} and 10^{15} tons in weight.

The limits of the biosphere are set by the ecological valencies of living systems. The upper limit of the biosphere coincides with the ozone layer, which absorbs a great amount of ultra-violet radiation. In the polar regions, life occurs only up to a height of 8 km in the atmosphere but in the equatorial zone it extends to a height of 18 km. Bacteria and the spores of moulds are regularly found at a height of 11 km and condors have been observed at 7 km.

The atmosphere is a mixture of gases held in place by the earth's gravity. The most important substances making up the atmosphere are as follows, measured by volume: Nitrogen = 78.1%, Oxygen = 20.94%; (According to measurements of the American National Bureau of Standards, the oxygen content of the atmosphere has remained essentially unchanged since 1910. It varies between 20.945% and 20.952%); Argon = 0.934%, Carbon dioxide = 0.03% (According to results of the Stanford Research Institute only 1.3% of the CO_2 of the atmosphere arises from artificial combustion processes. Since the beginning of the century the CO_2 concentration has risen at the rate of about 0.7 ppm per year (EBERAN EBERHORST 1972)); Hydrogen = 0.01%; Neon = 0.0018%; Helium = 0.0005%; Krypton = 0.0001%; Xenon = 0.000009%; Water vapour, variable.

The atmosphere can be divided into: 1. Troposphere; this is the layer, up to 20 km thick, with very strong horizontal and vertical mixing and containing water vapour. It is therefore the layer in which meteorological events occur. In some weather conditions the troposphere contains sharply delimited layers of air in which temperature increases with height, instead of decreasing (temperature inversions); 2. Stratosphere: This extends up to 50 km above the earth's surface with a temperature of −50°C to −85°C at the base; 3. Mesosphere; 4. Sonosphere; 5. Exosphere.

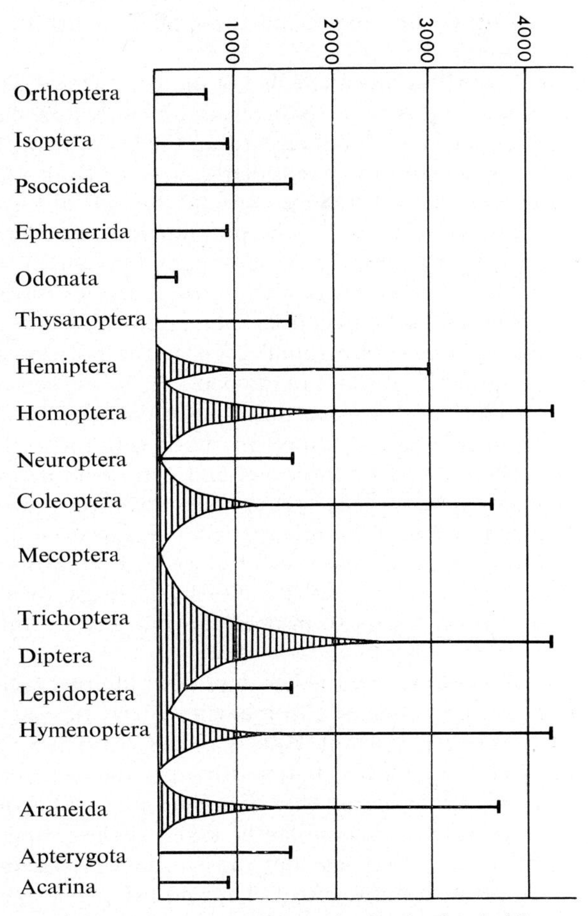

Fig. 5. Vertical distribution of insects, spiders and mites in the air (GLICK 1939).

Studies of aerial plankton (e.g. GLICK 1939) show that the atmosphere does not contain life everywhere, and that the species constitution varies considerably.

Most life requires free oxygen. Some plants and animals, on the other hand, can live without it, since they can obtain sufficient energy by fermentative breakdown of a richly available food supply. Many organisms are highly resistant to carbon dioxide, hydrogen sulphide and ammonia. Examples are animals that live on putrescent matter such as nematodes, gamasiform mites and Collembola (cf. MOURSI 1962).

Gases dissolved in water have a more obvious limiting effect than atmospheric gases. Dissolved oxygen may thus act as a limiting minimum factor, though of course different organisms behave differently with regard to it. The use of saprobes for evaluating water depends on the graded ability to use available oxygen (KOLKWITZ 1950).

The hydrosphere, like the atmosphere, is not uniformly filled with life. In the depths of the Black Sea, and in many Norwegian fjords that are closed at depth by a bar, the rich production of H_2S has used up the O_2 present. In the middle depths of tropical seas, where vertical circulation is very weak, there is a great lack of O_2. In summer many lakes are without oxygen at depth. In rivers downstream of big towns O_2 is used up by putrefaction caused by contamination by sewage. Thus the Thames, which above London contained 7.4 cm^3 O_2 dissolved oxygen per litre (HESSE 1924), only carried 0.25 cm^3/litre downstream of the city. Green phytoplankton occurs in the sea down to a depth of 400 m; and heterotrophous organisms are found even at the bottom of the Pacific deep-sea trenches, at a depth of more than 9000 m. The oxygen important for life processes can reach these places by diffusion or convection.

The lithosphere is in general only uniformly filled with life in the uppermost soil layer (= pedosphere). Apart from caves and petroleum deposits the solid earth is in general only inhabited by soil animals down to 5 m depth. Life is absent from actual solid rock. In petroleum deposits, however, anaerobic bacteria extend down to 4000 m (KUZNETZOV, IVANOV & LJALIKOYA 1962). Earthworms occur down to 8 m in the southern Urals (KÜHNELT 1970) and termite burrows are found down to 25 or 50 m deep in Madagascan forests (ERHART 1956).

Besides oxygen, temperature is a further important limiting factor. Since the course of physiological processes in all organisms obeys the van 't Hoff rule, ambient temperature is very important (HARDY 1972, COLINVAUX 1973, COLLIER, COX, JOHNSON & MILLER 1973). It is noteworthy that the usual limiting values are more restricted than the variation that occurs on earth. The lower temperature limit for poikilothermous animals can be taken at the temperature of 0°C to −2.5°C as found in the deep sea, but most organisms cannot survive in permanently low temperatures of this sort. Thus no fish live in the deep waters of the Hudson Bay, where the temperature in summer does not rise above 1.8°C (HUNTSMAN 1943). The scallop *Pecten groenlandicus* lives at a depth of 25 m off the coast of Greenland. The layers of water that it inhabits are relatively poor in food and it therefore tries to penetrate into shallower water. But as soon as it swims over the 0° line its metabolic rate, measured by oxygen consumption, rises so strongly that the species cannot obtain sufficient food for its needs. It is, therefore, forced by its metabolic physiology to live at the boundary between the superficial layer of food-rich water and the colder, deeper layer.

Similar adaptations also exist in terrestial animals (e.g. TISCHLER 1955, SCHWERDTFEGER 1968). The snow-fauna can live on snow or ice as a substrate and mainly feeds on organic remains blown on to the snow – the so-called cryoconite. Among the snow fauna is the glacier flea, a primitive insect of the order Collembola. In the Alps this is represented by species of the genus *Isotoma*; these overwinter on the surface of the old snow, underneath the new snow, and feed on the needles of conifers. Low heat conduction and low temperature, strong reflection of radiation and low storage of diurnal heat are ecological conditions in snow regions to which the snow fauna needs to be adapted (e.g. FRANZ 1969, KÜHNELT 1969). The snow and ice areas of the arctic and antarctic and of the oreal regions of high mountains are characterised by several species of diatoms, Desmidiaceae, Chlorophyceae and Cyanophyceae which are known as snow algae. These can be so abundant that their red or violet pigments redden the snow, producing 'blood snow' that is visible at some distance. The alga *Haematococcus nivalis* is an extreme case; it imports a red colour to the firn of the Alps and polar regions. It needs a temperature of 0°C to flourish and ceases to grow above a temperature of +4°C. The frost limit is very important for vegetational geography since for a large number of plants 0°C represents the extreme limit of development.

Homoiothermous animals can live in even lower temperatures. DÖRRE (1926) reported the presence of the nest of a house mouse, with seven living young, near the cold ducting of a Berlin factory, with a permanent temperature of −11°C to −12°C. In the cold stores of the city of Hamburg mice used to live on frozen bacon and made their nests in frozen meat at a temperature of −6°C (MOHR & DUNKER 1930).

The upper limit of temperature, like the lower limit, is of decisive importance. The rhizopod *Hyalodiscus* lives in hot springs at a temperature of 54°C (BRUES 1928). The water snail *Bithynia therminalis* occurs in the thermal springs of Rome at 53°C. The ostracod crustacean *Cypris balnearia* and the midge *Dasyhelea terna* tolerate a temperature of 51°C. In hot springs the blue-green alga *Synechococcus lividus* occurs at 74°C (PEARY & CASTENHOLZ 1964). Cyanophyceae and bacteria can still be found at a temperature of almost 90°C. Fungal spores and microbes have been exposed to temperatures of between 140°C and 180°C without being killed (SCHMIDT 1969, BROCK 1967, EDWARDS & GARROD 1972, ZIEGLER 1969). SAUSSURE found eels at Aix at 46°C. *Leuciscus thermalis* is supposed to exist in springs at Trincomalee at 50°C (quoted from HESSE, 1924). The amphibia *Hyla raniceps*, *Bufo paracnemis*, *Leptodactylus ocellatus*, *L. pentadactylus* and *Pseudis bolbodactyla* spawn at 38°C in springs and brooks in the area of Pousada do Rio Quente (Goias, Brazil, MÜLLER 1971).

The resistant phases of various plants and animals have an even greater ability to withstand cold and heat, considerably exceeding the variations in temperature which now exist upon earth. The cysts of the tardigrade *Macro-*

biotus hufelandi, which has a world-wide distribution, will revive in water when they have previously been: i) left 20 months in liquid air, at −190°C to −200°C; ii) or left 8½ hours in liquid helium at −272°C; iii) exposed for 10 hours to a temperature of 60 to 65°C; iv) or exposed for 1 hour to a temperature of 92°C. On the other hand, they will tolerate a temperature of 100°C only for 15 minutes, and in this case 30% of them are killed. A temperature of 103°C is always fatal. The cysts will revive in water after being left for 60 days in a desiccator at an air humidity of 8% (KAESTNER 1965).

Under experimental conditions the African midge *Polypedilum vanderplanki* can survive, after drying out, a warming to 100°C or cooling to −196°C. Bacteria can also be cultivated after spending up to half a year in liquid air at −190°C (SCHMIDT 1969).

Certain substrates of extreme chemical composition sometimes form abiotic enclaves in the biosphere. It is nonetheless striking that some areas whose chemical composition is highly unusual can sometimes be conquered by particular species, using special adaptations. Thus the ephydrid *Psilopa petrolei* inhabits oil pools. This 'petroleum fly' has been recorded on animals or the remains of animals that have fallen into these pools.

At fumaroles the earth produces abundant CO_2 which, being denser, drives away all air at ground level; an example is the Grotta del Cane at Pozzuoli. At such places all life is excluded. At the mofettes on the eastern bank of the Laacher See (Eifel) it is fairly common to find the bodies of little birds and mammals, such as finches and mice, which had blundered into the CO_2 atmosphere while looking for food and died as a result.

The limiting effect and significance of pH on various organisms has been discussed exhaustively by BROCK 1969 and EDWARDS & GARROD (1972).

In summary, the biosphere includes firstly, the hydrosphere which is divided into individual water bodies; secondly, a relatively thin layer of the lithosphere known as the pedosphere, as well as caves and petroleum deposits; and thirdly, the lower layers of the atmosphere. There are, however, certain abiotic enclaves such as volcanoes, ice, abiotic regions at the bottoms of lakes and sea basins, and abiotic regions caused by extremes of pH.

THE DISTRIBUTION AREA AND THEIR DYNAMIC

All zoogeographical studies begin from knowledge of an area of distribution. Before a geographical range can be explained it is necessary to describe its spatial distribution precisely. This requirement for well proven limits of range can be approximately satisfied only for the vertebrates even in such well worked regions as Europe and North America. The limits of the ranges of species and subspecies can only be given with greater or lesser probability unless they are 'fixed' by natural barriers to distribution such as water bodies, mountains or competing species.

Species are groups of populations made up of mutually fertile individuals. They represent the largest possible potential reproductive communities. They are reproductively isolated from other such reproductive communities and can be sympatrically distributed with them – i.e. exist in the same area – without losing their identity (HUXLEY 1940, MAYR 1942). By contrast with monotypic species, polytypic species have a range consisting of at least two subspecies ranges. **Subspecies**, or geographical races, are populations of a polytypic species that are confined to a particular area. They can be distinguished from other allopatric populations of the same species – i.e. populations with which they are mutually exclusive in range – by a characteristic feature in 70% of individuals. Subspecies, unlike species, are in general fertile with each other, so that contact between the ranges of subspecies usually produces zones of hybridisation. For reasons of evolutionary genetics, subspecific populations need to be distributed allopatrically to preserve their identity. Subspeciation can result from differing selection-pressures in different regions or from spatial isolation (cf. Dispersal Centres by MÜLLER 1973). As a result of a long-continued phase of isolation, subspeciation can lead to speciation, so that subspecies can be seen as species in statu nascendi.

Substrate races are a particularly interesting type of subspecifically differentiated population. Such races are usually subspecifically differentiated populations of a polytypic species which are linked to a special type of soil. The concentration of a particular phenotype at a particular place can be interpreted in this connection as due to selection (MAYR 1967). The production of a local coloration congruent to a particular type of substrate is a phenomenon that has been proved in a great variety of animal and plant groups. Examples are rock plants of the genus *Lithops*, the desert lizard *Eremias undata gaerdesi*, the Viperid *Cerastes cerastes*, the elephant-shrew *Elephantalus intufi namibensis* and

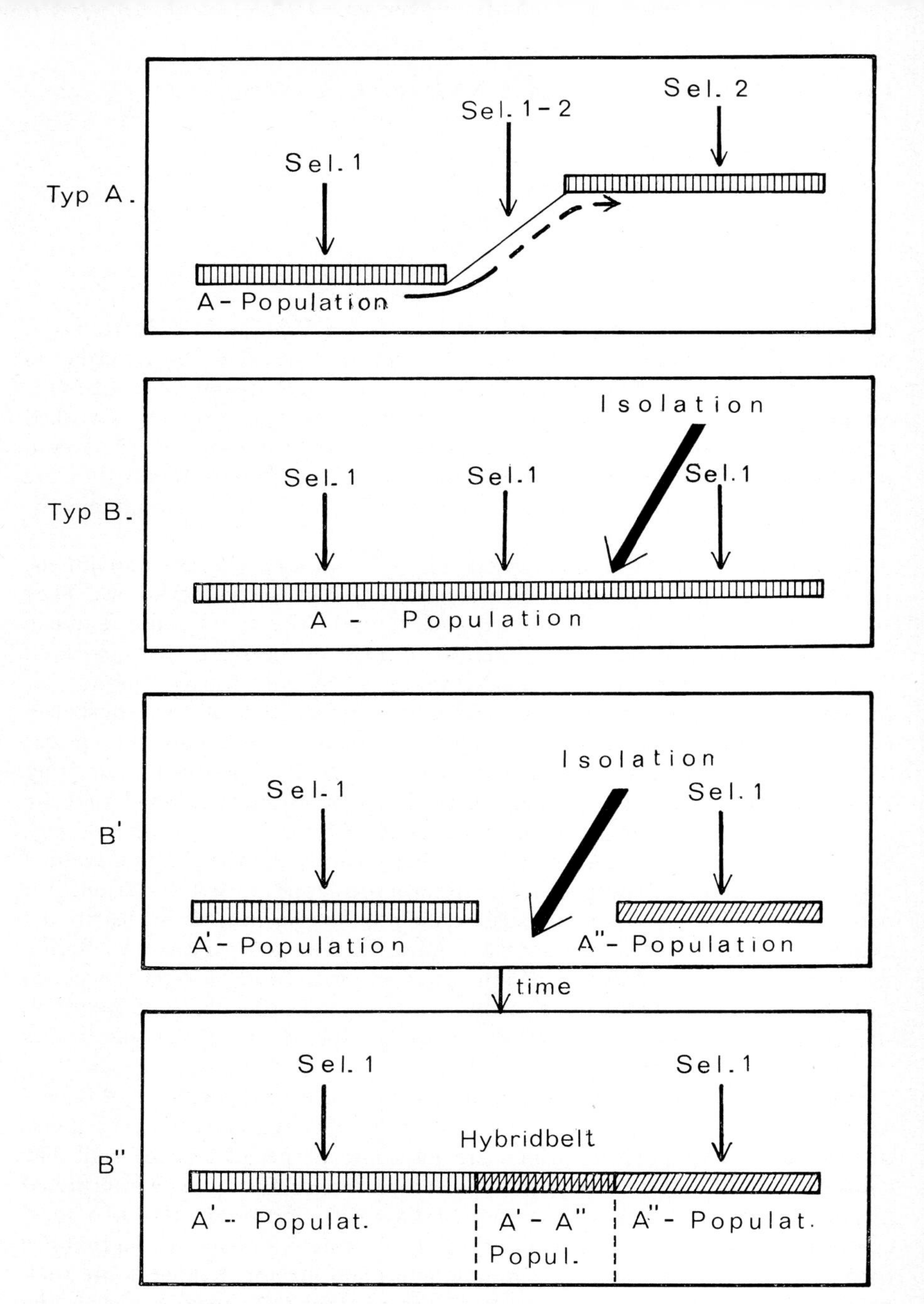

Fig. 6. The two general models of subspeciation (A and B); In A = differentiation as a result of migration and differential selection; In B differentiation as a result of separation (after MÜLLER 1973).

the rodent *Gerbillus gerbillus leucanthus*. LEWIS (1949) described substrate races in the lizards of New Mexico. BENSON (1933), BLAIR (1943) HOFFMEISTER (1956) and BAKER (1960) have studied some remarkable races of particular species living on lava. And NIETHAMMER (1959) and VAURIE (1951) have concerned themselves with the substrate races of certain species of birds.

Also distributed allopatrically are semispecies, belonging to a **superspecies complex.** Superspecies are groups of monophyletic origin made up of essentially allopatrically distributed species (semispecies). These latter are morphologically too different from each other and their fertility with each other is too limited for them to be considered as belonging all to a single species (KEAST 1961, MAYR 1967).

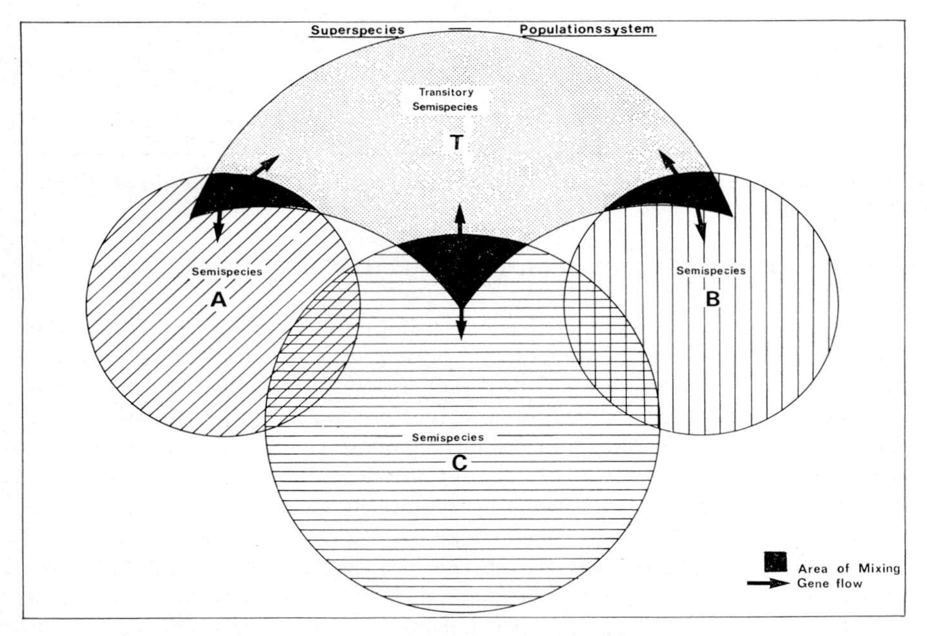

Fig. 7. The Superspecies population system (after SPERLICH 1973).

If a subspecies A is found on the east bank of a river, and a subspecies B on its west bank, it is highly probable that the river represents the limit of range between them. For separate species, which can exist side-by-side in the same region, such a conclusion is much more difficult to justify.

Natural regions of distribution are nowadays threatened from all sides. For this reason it is now more urgent than ever to work out geographical ranges as quickly as possible (BROWN 1964, BUHL 1969, CROWE 1966, FITZSIMONS 1962, Flora Europaea 1972, JUNGBLUTH 1972, HEATH and LECLERCQ 1970, HEATH

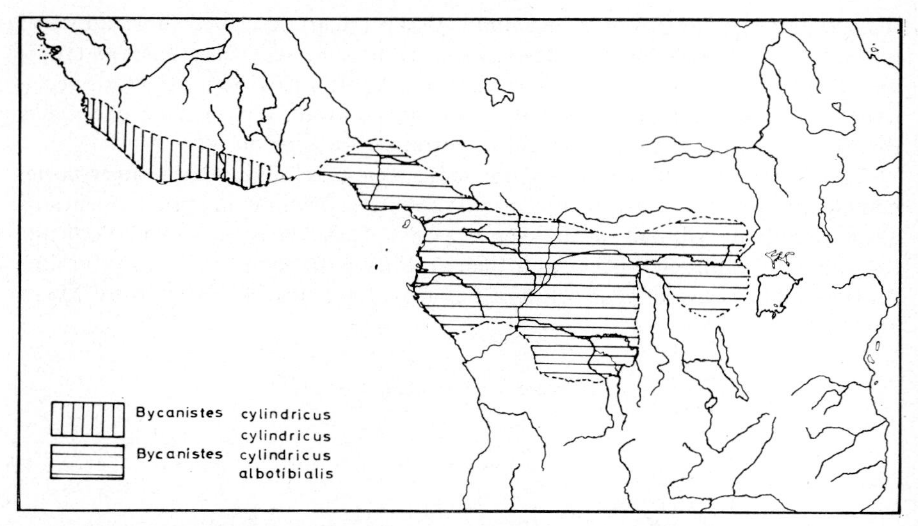

Fig. 8. Allopatrically distribution of the subspecies *cylindricus* and *albotibialis* of the rain-forest species *Bycanistes cylindricus.*

1971, LLOYD 1962, MÜLLER 1972, 1973, NIKLFELD 1972, PERRING 1963, 1965, PITELKA 1951, SOPER 1966).

Only after all the facts of distribution are known can studies be carried out to explain origins and throw light on the present-day dynamics of geographical ranges. In this connection it is important to stress that a particular piece of distribution data which confirms the presence of a species in a particular region, represents very incomplete information. In many cases, as for example with parasites, such information is of very subordinate importance. Setting out the facts of distribution means setting out the localities and also the environments in which organisms live in particular regions.

The true **geographical range of a species** of animals implies in the first place only its breeding range. This is the part of its region of distribution in which the species can reproduce itself permanently, without continual recruitment from outside (DE LATTIN 1967). This restricted sense of the concept of geographical range is necessary for several reasons. On the one hand it clearly separates the ranges of habitation, of reproduction, and of migration of a species, or subspecies. On the other, the area in which the handing down of group characteristics is assured by reproduction is thus assigned the first-rank importance which it truly possesses in the evolutionary history of a population.

Some species can readily be crossed with each other in the laboratory but only preserve their characteristic features because they visit the same breeding range at different times. They are isolated not geographically but in time. In such cases the sizes of the ranges can be very different in the different groups.

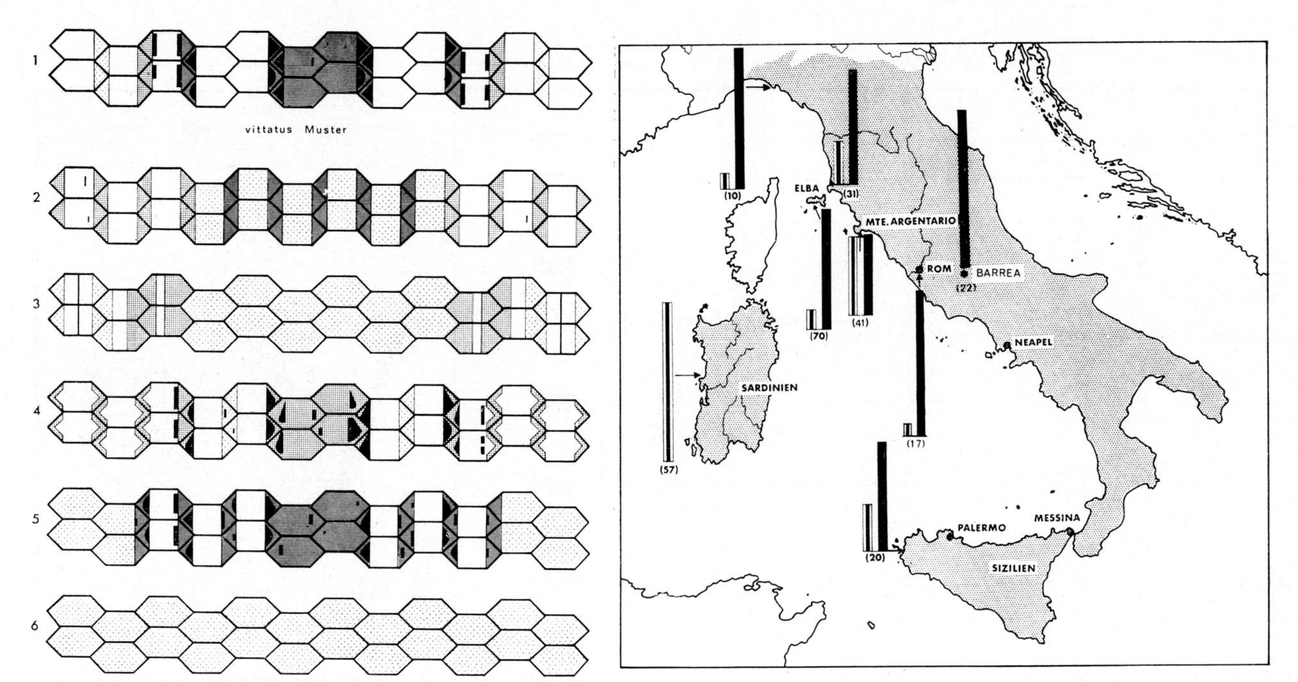

Fig. 9. Types of pattern in Italian *Chalcides chalcides* and the relative abundance of vittatus-pattern (striped column) in Italian populations. The numbers are the specimens studied.

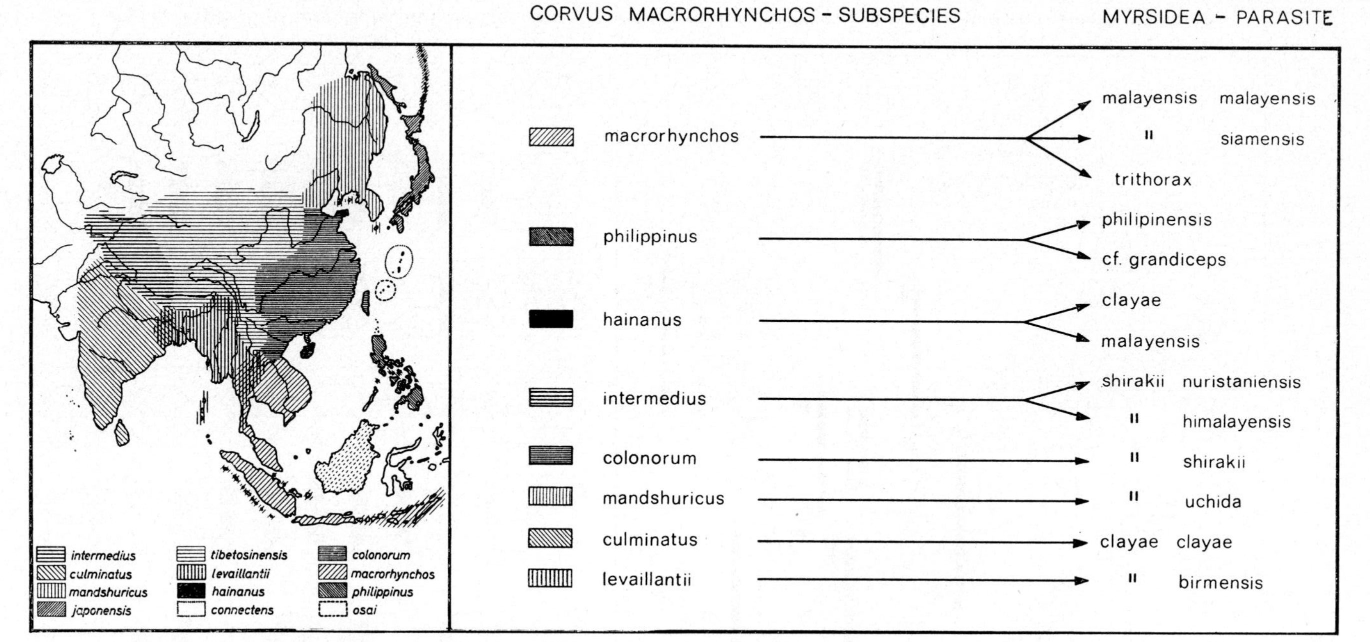

Fig. 10. Coincidence of range between host (*Corvus macrorhynchus*) and parasite (Myrsidea), (after KLOCKENHOFF 1969).

Their size depends on the ecological valency, the possibilities of movement and dispersal, the history of distribution and the geographical position of the area of origin of a species. On the other hand a general connection between the size of the range and the phylogenetic age of a species, as implied by the 'age and area rule' of Willis, only exists in the rarest cases. Many species and genera occur in all parts of the earth as **cosmopolitan groups**, but are nevertheless of very different phylogenetic ages. They show that no simple connection between age and range exists. Cosmopolitan groups, because of their ecological valency, powers of dispersal or their close association with man – as domestic animals, parasites or easily introduced species – are not confined to any particular zoogeographical realm. However, there are no species which exist equally in the sea, in fresh water and on land. Examples of cosmopolitan animal species are: the barn owl (*Tyto alba*); the osprey (*Pandion haliaetus*); the grey heron (*Ardea cinerea*); the peregrine falcon (*Falco peregrinus*); the painted lady butterfly (*Vanessa cardui*) (which inhabits all the continents except South America), the butterflies *Plutella maculipennis* and *Nonnophila noctuella*; and the tardigrade *Macrobiotus hufelandi* whose resistant phases can withstand the most extreme external conditions. There are also numerous cosmopolitan species of plants such as the annual meadow grass (*Poa annua*); the great plantain (*Plantago major*); the dandelion (*Taraxacum officinale*); the fat hen (*Chenopodium album*); the common reed (*Phragmites communis*); and the stinging nettle (*Urtica dioica*).

Although cosmopolitan groups are distributed world-wide, they nonetheless possess certain local requirements. This distinguishes them from **ubiquists** which occur everywhere, have no special local requirements and can survive and reproduce under very unfavourable conditions. Many species of bacteria and algae can be regarded as ubiquists (WOLFENBARGER 1946).

In contrast to cosmopolitan and ubiquitious species are those species and populations which at present only occur in a very small part of the earth, as relicts. Such **relicts** are organisms that once had a wider distribution but whose ranges have been diminished, split up or shifted because of a change in environmental conditions. As a result they can now survive only in particularly favourable places. Relicts can be of various ages – Tertiary, Pleistocene, Holocene, etc.

Tertiary relicts are taxa which have survived to the present at least since the Pliocene in a particularly favourable place, or refuge. The concept of the Tertiary relict can only be applied to cases where the preglacial, relict character of a population can be proved both from the phylogenetic aspect (constancy of features) and the biogeographical aspect (constancy of locality). It is above all appropriate for biotopes that were little influenced by Pleistocene climatic oscillations – such as old lakes, bodies of ground water, thermal springs and caves – and for plant and animal species with a slow rate of evolution.

Glacial relicts are animal or plant species which have existed in their recent localities since the Würm glaciation as remains of stenothermal biotas adapted to a colder climate. Examples of these among central European animals are the flatworm *Planaria alpina* that lives in the rhitron (uppermost reaches) of the rivers of the Central European mountain ranges and belongs to the fauna of the springs. Other examples are the cave animals *Onychiurus sibiricus* (Apterygota), *Pseudosinella alba* (Apterygota) and *Choleva septemtrionis holsatica* (Coleoptera). Glacial relicts among plants include the alpine *Lycopodium* (*L. alpina*) and the dwarf birch *Betula nana* on the Brocken of southern Germany. Even some of the invertebrate species adapted to the treeless, alpine high mountain region, such as ground beetles of the genus *Trechus*, may have survived the Würm glaciation in mountain regions that remained free of ice (so-called 'massifs de refuge') or on individual mountain summits which stuck out of the glacial ice as nunataks (HOLDHAUS 1954, BESUCHET 1968, NADIG 1968). The newer results of glaciological researches in Scandinavia, concerning the course and duration of the Würm glacial period, cast doubt on the existence of glacial relicts in this area. The question of small refuges in the Würm glaciation in Scandinavia must therefore be reconsidered (LINDROTH 1939, 1969, STOP-BOWITZ 1969). Thus LINDROTH (1969) was able to show that certain Carabid beetles that had formerly been taken as glacial relicts had only survived from the Younger Dryas Period or at most Würm II, which according to SHOTTON (1967) and LUNDQUIST (1964, 1967) was separated from Würm I by an interstadial. If so, these species are not strictly glacial relicts. On the other hand glacial relicts may have been able to survive to the present in isolated lakes of the northern Holarctic region (SEGERSTRALE 1966). Among the crustaceans, examples are the species: *Mysis relicta*, *Mesidotea entomon*, *Gammaracanthus lacustris*, *Pontoporeia affinis*, *Pallasea quadrispinosa* and *Limnocalanus macrurus-grimaldii*. These animals of the Baltic region represent a relict fresh-water fauna of the Holarctic, whose closest relatives live in the sea or can be derived from marine groups. Populations of such Baltic animals occur in the Baltic Sea and also in many Scandinavian, British, Finnish, North Russian, Siberian and North American lakes. The Fennoscandian populations were directly connected with the Baltic Sea during the post-glacial period and only became isolated by the retreat of that sea. This interpretation was first put forward by EKMAN (1940) and thus became known as the Ekman theory. According to it the occurrence of animals of the Baltic region in lakes and in the Baltic sea represent relict populations. Such species originally had a marine arctic range and immigrated during the cool Yoldia period, or slightly before, by way of a marine connection from the White Sea, through Lakes Onega and Ladoga to the Yoldia Sea. As the level of water changed they penetrated into the various Fennoscandian lakes and remained as relicts after the water level fell. The Ekman theory was further developed by SEGERSTRALE (1957, 1962).

Compared with glacial relicts, xerothermal relicts are generally considerably younger. These are thermophile populations which once had a wider distribution. Their geographical ranges have contracted, split up or shifted as the result

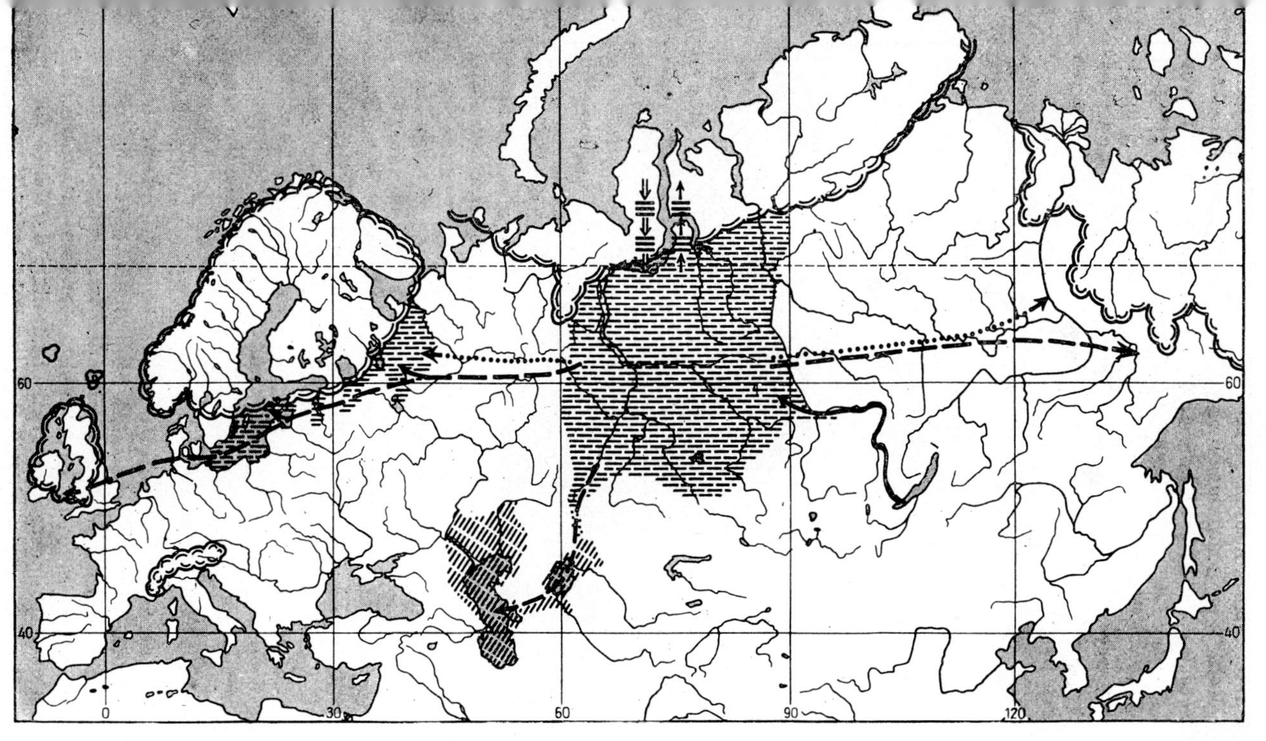

Glaciers (The mountain glaciations of southerly regions are not included, except for those of the Alps.

Ice-dammed lakes.

Aralo-Caspian Lake.

Region through which the dammed-up West Siberian Ice Lake temporarily drained into the Aralo-Caspian basin.

Symbol for the penetration of marine brackishwater animals into the Siberian Ice Lake during a glacial advance.

Migration routes westwards, eastwards and southwards of species of originally marine origin.

Migration routes westwards and eastwards of species of originally Baikalian origin.

Migration route by which Baikalian species invaded the West Siberian Ice Lake, and conversely.

Symbol for the retreat of animals of the Siberian Ice Lake into marine brackish water during a glacial recession.

Fig. 11. The late glacial and postglacial ice-dammed lake and its significance for the distribution of glacial relicts in the Palaearctic region (based on SEGERSTRALE 1957, from DE LATTIN 1967).

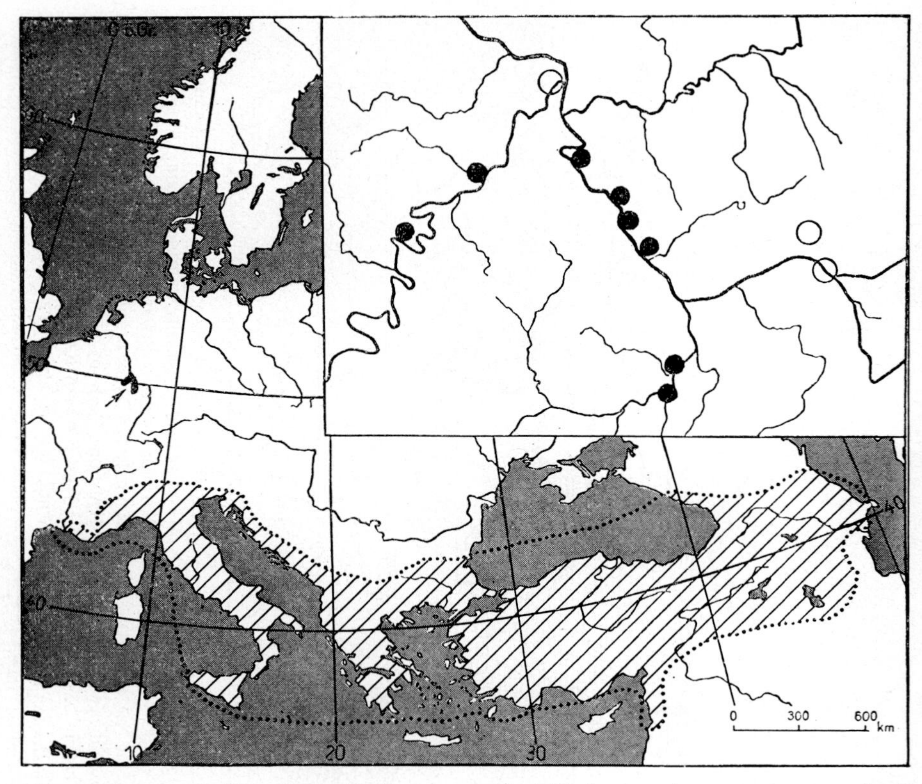

Fig. 12. The range of the Ponto-Mediterranean Noctuid *Ammoconia senex*. An example of the distribution of a thermophile species with an isolated postglacial relict occurence (black) in the Middle Rhine region-subspecies *mediorhenana*.

of worsening climate. They have therefore been able to survive only in islands of warmth where the climate is specially favourable. Xerothermal relicts have been particularly well studied in the Moselle and Middle Rhine areas (DE LATTIN 1967, MÜLLER 1971, WARNECKE 1927). The xerothermal relicts of the Moselle area reached their present locations during the climatic optimum (5000–1000 B.C.), probably via the Moselle and Rhine trenches (MÜLLER 1971), which served as migration routes. In the succeeding 'Beech Age', which began with a drop in temperature and an expansion of the beech forests, the post-glacial migration routes were interrupted and the Moselle and Middle Rhine populations separated from their parent populations in the Mediterranean area. The brief phase of isolation (about 3000 years) had led in many cases to the development of clearly distinguishable subspecies within the isolated populations (DE LATTIN 1967). Thus the Apollo butterfly *Parnassius apollo* occurs at Winningen (Mosel) as a subspecies endemic to the area (*P.a. vinningensis*). Many relicts

Fig. 13. The biotope of thermophile species in the Saar – Mosella area; Hammelsberg bei Perl, June 1970.

have in common that they exist as tiny, isolated populations far outside the main range of the species.

Particularly important for historical zoogeography are those ranges which are split into several smaller subranges. These are known as **disjunct ranges**. Populations with a disjunct distribution are lacking, for ecological or historical reasons, or both, from the regions separating the sub-ranges. The splitting up of originally uniform populations is an essential precondition for geographical species formation. There exists a great variety of types of disjunct range. One much discussed type is the bipolar distribution. Taxa with bipolar ranges are absent from the tropics and exist only in the higher latitudes of the northern and southern hemispheres (BERG 1933, DE LATTIN 1967). Among plants, *Viola*, *Papaver*, *Empetrium* and the Fagaceae belong to this type of distribution along with some 60 other genera. A similar type of distribution occurs in the pelagic and benthic fauna of the sea, but it has been shown that certain apparent bipolar distribution patterns in fact arise by equatorial submergence. This is a vertical shift in range of cold-loving marine animals in the tropical regions, thus avoiding the surface waters that are preferred in higher latitudes. In this way the species crosses the equator by way of the deeper and cooler layers of water. Thus the copepod *Rhincalanus nasutus* inhabits the surface waters of the Atlantic north of 40°N and south of 30°S. Between 10°N and 10°S it only occurs deeper

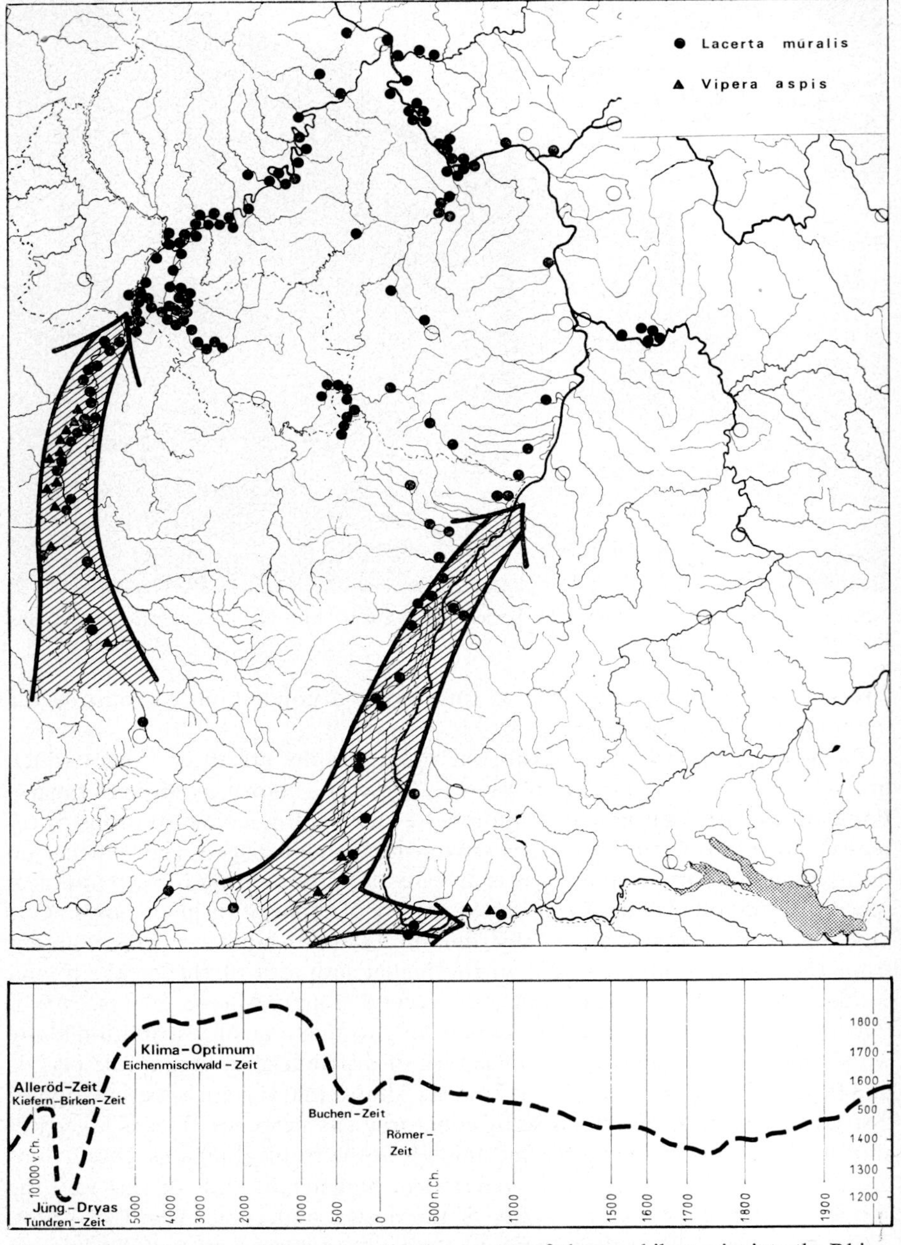

Fig. 14. The treferred postglacial immigration routes of thermophile species into the Rhine-Moselle area (from MÜLLER 1971).

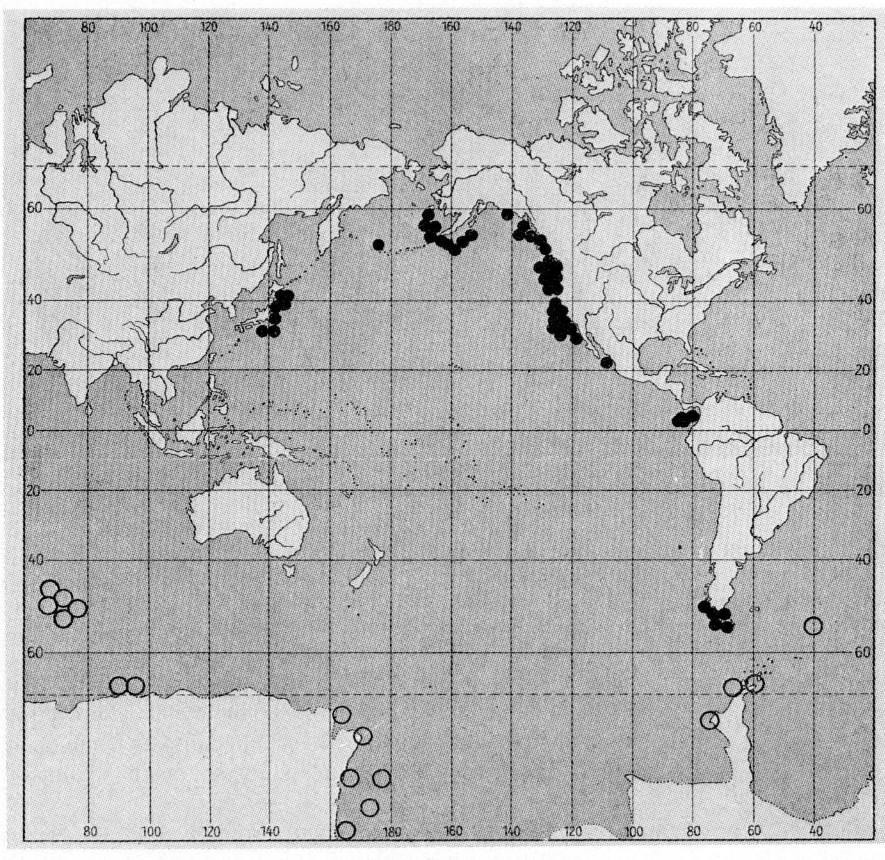

Fig. 18. Bipolar distribution of Promachocrinus (after EKMAN 1935; from DE LATTIN 1967).

Table I: Palaearctic and Aethiopian migratory birds (after MOREAU 1972); The number of species flying to Africa (no brackets) together with separate subspecies (brackets).

	West Palaearctic	Central Palaearctic	East Palaearctic
Warblers	26+(2)	20+(1)	2
Other passeriforms	36+(5)	32+(5)	6+(1)
Birds of prey	19+(5)	5	3
Water birds	35+(28)	12+(1)	1
Other non-passeriforms	21	13	2

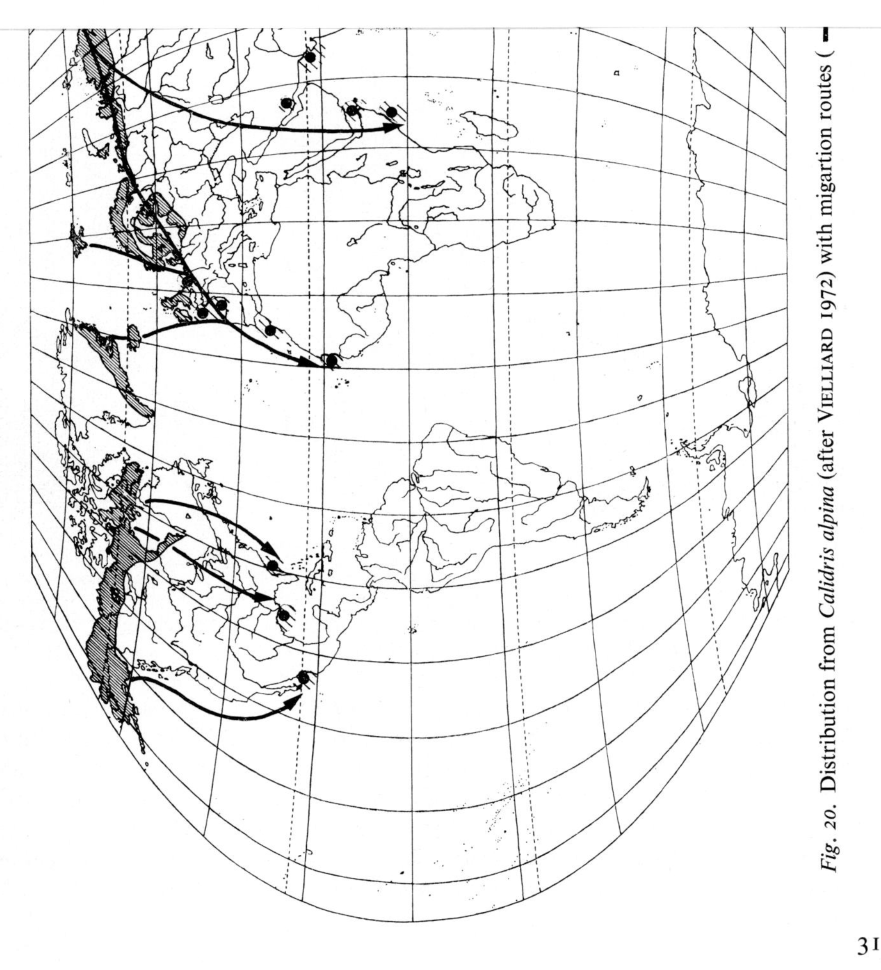

Fig. 20. Distribution from *Calidris alpina* (after VIELLIARD 1972) with migartion routes (▬

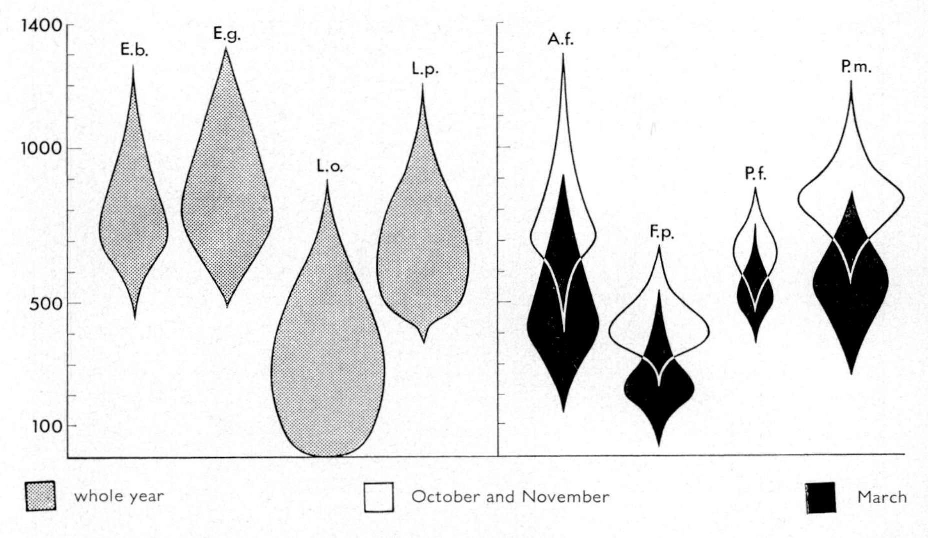

Fig. 19. Seasonal change in vertical distribution of the birds *Amazona farinosa*, *Forpus passerinus*, *Pyrrhura frontalis* and *Pionus maxiliani* on the island São Sebastião is correlated with the breeding season in October and November. The amphibia are *Eleutherodactylus binotatus*, *E. guntheri*, *Leptodactylus ocellatus* and *L. pentadactylus* (after MÜLLER 1973).

There have certainly been bird migrations since the early Tertiary. But the migration routes of present-day migratory birds are nevertheless extraordinarily young. 'All such adaptations must be the product of evolution in something like the last 10,000 years, a conclusion shattering to much current evolution theory'.

Both in the Norwegian lemming (*Lemmus lemmus*) and also in the wood lemming (*Myopus schisticolor*) there is a marked seasonal shift in biotope, which is strongly conditioned by the seasonal changes in the most important food plants (mosses, grasses). The long-distance migrations of the Norwegian lem-

the pale clouded yellow butterfly (*Colias hyale*), the death's head hawk-moth (*Acherontia atropos*) and the convolvulus hawk-moth (*Agrius convolvuli*).

Irregular migrations, which can happen from very various causes, are also carried out by dragonflies (e.g. *Sympetrum fonscolombei*, *Hemianax ephippiger*, *Aeschna affinis*) and millipedes. At irregular intervals there are mass movements of Pallas' sand grouse (*Syrrhaptes paradoxus*), the nutcracker (*Nucifraga caryocatactes macrorhynchus*) and the crossbill (*Loxia*), which lead to invasions of Central Europe. In invasion years huge flocks of the nutcracker leave Siberia in a westward direction. These movements are mainly caused by specially

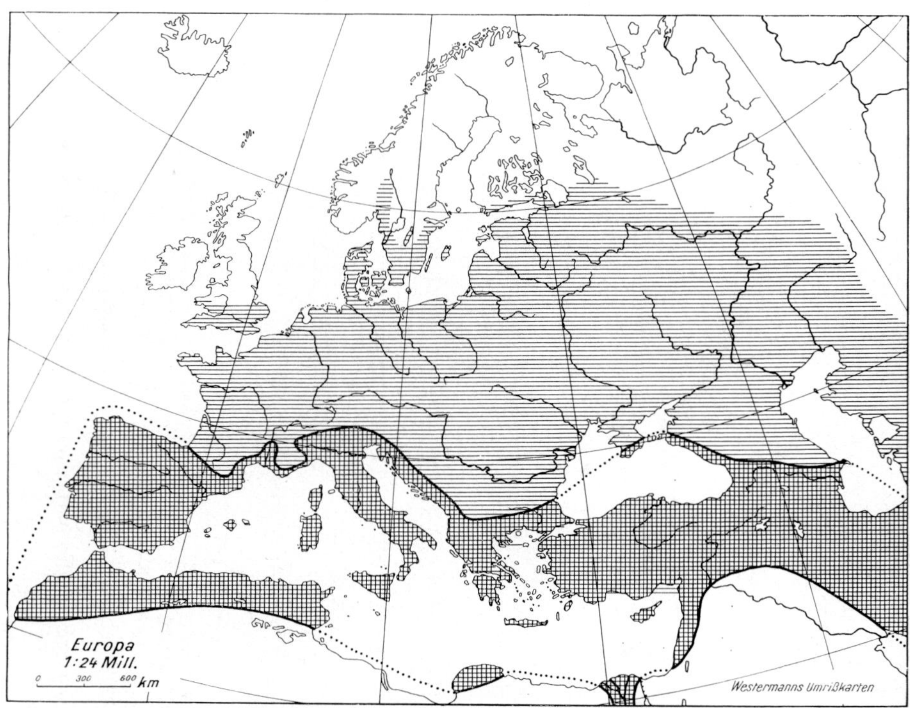

Fig. 21. Breeding range (cross-hatching) and migratory range (vertical hatching) of the Holomediterranean migratory butterfly *Colias croceus* (after DE LATTIN 1967).

abundant seed production in the conifers of the invaded regions and are also conditioned by the population density. Invasion years for the Siberian nutcracker into Europe were: 1753, 1754, 1760, 1793, 1802, 1804, 1814, 1821, 1827, 1836, 1844, 1849, 1856, 1868, 1878, 1883, 1885, 1888, 1893, 1895, 1899, 1900, 1907, 1911, 1913, 1917, 1933, 1941, 1947, 1954, 1961 and 1968.

Irregular rainfall-ralationships lead to population explosions and invasions of migratory locusts in the savannas and steppes (cf. savanna biome).

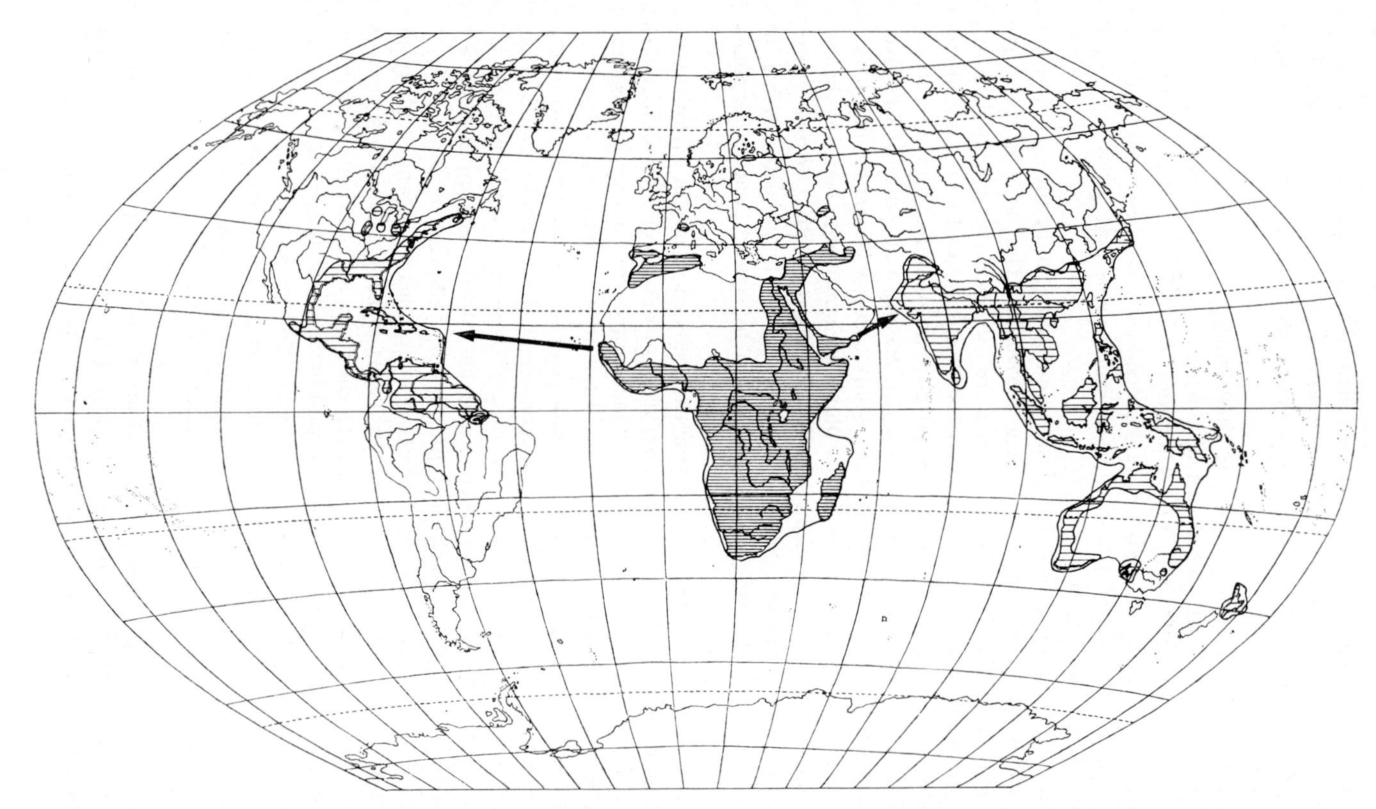

Fig. 22. The range of *Ardeola ibis*. Original range, black; range due to expansion in historical times, shaded. (after SCHÜZ & KUHK 1970).

It is a general characteristic of migrations that the true reproductive area or breeding range does not essentially expand except in very rare cases. This is quite different from changes in range which depend on 'passive' dispersal mechanisms.

Many plants and animals increase their ranges by passive mechanisms of dispersal. Two types of such dispersal can be distinguished in plants (cf. e.g. RIDLEY 1930), i.e. autochory and allochory. Autochory is the formation of suckers, the spreading of seeds or the dispersal caused when a part falls under its own weight (barochory). Allochory is when seeds and fruits are passively distributed by other means.

Among steppe plants it is possible for whole bushes to be torn out and blown before the wind as 'steppe-witches' (e.g. *Serratula*). The same is true for the spherical thalli of the manna lichen (*Lecanora esculenta*), which can drift before the wind, rolling over the ground. Many very small, dust-like seeds and spores are distributed by air, including the spores of mosses and ferns, the basidiospores of mushrooms and toadstools, the conidia and ascospores of ascomycetes and the seeds of orchids and tobacco. Many seeds with special flight mechanisms are also distributed by air. Other plants are adapted to water as the means of transport. Buoyancy is achieved by air-carrying cells (*Alisma plantago-aquatica* or water plantain, *Sagittaria sagittifolia* or arrow head), by big intercellular spaces (*Caltha palustris* or kingcup) or by special floats (*Nymphaea* or water lily, and sedges).

It was already shown by DARWIN that the seeds and fruits of various types of plant could float for a long time in either fresh or salt water without losing their fertility. RODE (1913) came to similar conclusions. Particularly interesting are those plant seeds which are adapted for distribution by animals. These include the various types of burr which cling to the pelts of animals by means of hooks or barbs e.g. corn crowfoot (*Ranunculus arvensis*), the burdocks (*Arctium*), herb bennet (*Geum urbanum*). A particular form of these epizoochorous types is the grapple plant (*Harpagophytum procumbens*) that occurs in steppe and desert and is carried by catching on to the legt and fees of mammals.

Other seeds and fruits fix themselves on to animals by means of gluey slime and are distributed in this way (RIDLEY 1930). Synzoochorous distribution is achieved when animals carry off fruits to eat them undisturbed or in order to build up food stores. Examples of such animals are hamsters, squirrels, mice, jays, nutcrackers and woodpeckers (cf. HOLTMEIER 1966).

Myrmecochorous plants such as greater celandine (*Chelidonium majus*) and violets (*Viola*) carry brightly coloured, oil-rich bodies (elaiosomes) on their seeds. These are pulled to pieces with great enthusiasm by ants, which act as distributors.

Fruits with endozoochorous dispersal are usually obvious because of their attractive colour or their soft flesh. A hard shell protects the seed from being digested e.g. berries, large-stoned fruits such as plums, mistletoe, etc.

Passive distribution of animals occurs mainly by wind, by flowing water, or by being carried by other animals or by man.

In general it is only small and light animals that are suited to transport by air as aerial plankton. In the passive dispersal of insects air currents are particularly important. In general, the faster the currents are, the more insects they carry. This wind-drifted aerial plankton goes up to 4000 m high.

The spruce aphid *Cineropsis piceae* regularly occurs in Spitzbergen when the weather conditions are suitable. This is more than 1000 km distant from the nearest occurrence on the Kola peninsula. Descending winds over the land surface finally force the aerial plankton down in many regions. The result is almost daily and hourly a steady bombardment of the ground with little insects. Passive carriage in which wind and water are involved is said to be anemohydrochorous.

Anemohydrochory has been studied in depth by PALMEN (1944). He investigated the extensive insect windrow that commonly occurs on the south coast of Finland; the density of insects washed ashore in this windrow is about 4000 individuals/sq. metre. He was able to show that the windrow mainly consisted of insects from the Baltic coast opposite south Finland i.e. from about 100 km away. Some of the insects had survived for several days in salt water. The regularity of anemohydrochorous dispersal in this region is the reason for the Baltic immigration route of many Finnish insects.

In hydrochorous dispersal it is mostly aquatic animals that are carried, with floating material acting as the mode of transport (the rafting theory). Carriage by other animals (zoochory) applies especially to parasites and epizoans such as Mallophaga. Marsh and water birds can carry the eggs of aquatic insects or the spawn of other aquatic animals over great distances. A famous example is the 'wandering duck', carrying the fresh spawn of a fresh-water snail, which was shot by TRISTRAM in the Sahara, about 160 km from the nearest fresh water. WEIGOLD (1910) investigated the Chydorid fauna of the ponds of Saxony and discovered that the more a water body was visited by migrant birds, the richer it was in species of Chydorid. Nothing definite can be said about the origin of the range of a species unless its passive dispersal mechanisms are known.

Dispersal by man is mostly correlated with the expansion of particular peoples or with the principal routes of communication. Brown rats, house mice and house sparrows have become cosmopolitan through man's activities.

Many species of reptiles and amphibia have also been introduced. The big, South American giant toad *Bufo marinus* has intentionally been introduced into Cuba, Haiti, eastern Australia and New Guinea as an insect control measure. The little gecko *Hemidactylus mabouia* reached South America from Africa unintentionally with the first slave ships. Today it has become a 'domestic animal' in South America. The brown trout, which was originally confined to western Eurasia, has been introduced for economic reasons into North America, Chile, Argentina, South and East Africa, Madagascar, Australia and New

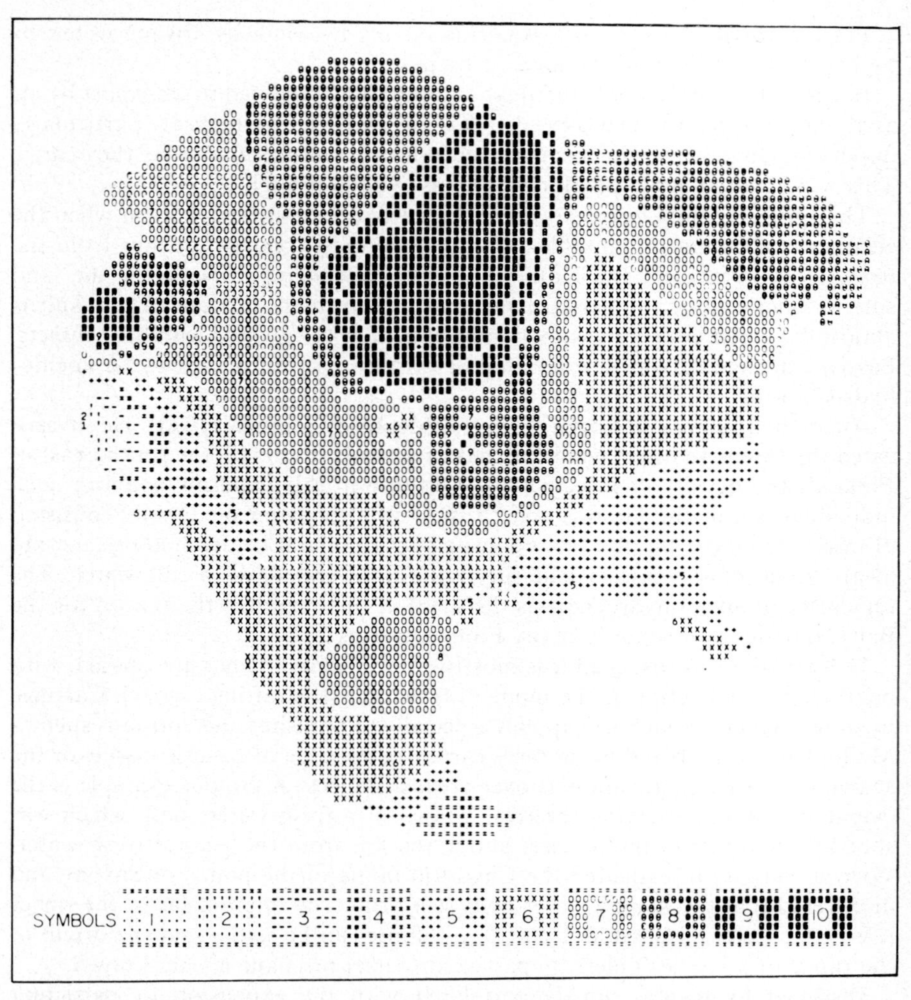

Fig. 23. Computer-generated map of values of principal component of variation in size of skeletal in male House Sparrows (*Passer domesticus*) from North America (from SELANDER 1971).

Zealand. The oyster (*Ostrea edulis*) has been distributed by man beyond its original range in the Mediterranean for the same reasons. Among European animals, the wild boar, the common hare and the starling have been carried to North America. Originally North American species that have been able to settle in Europe are the Colorado beetle, the musk rat, the grey squirrel and the raccoon.

Multiple introduction is known in many species. Thus the Algerian hedgehog

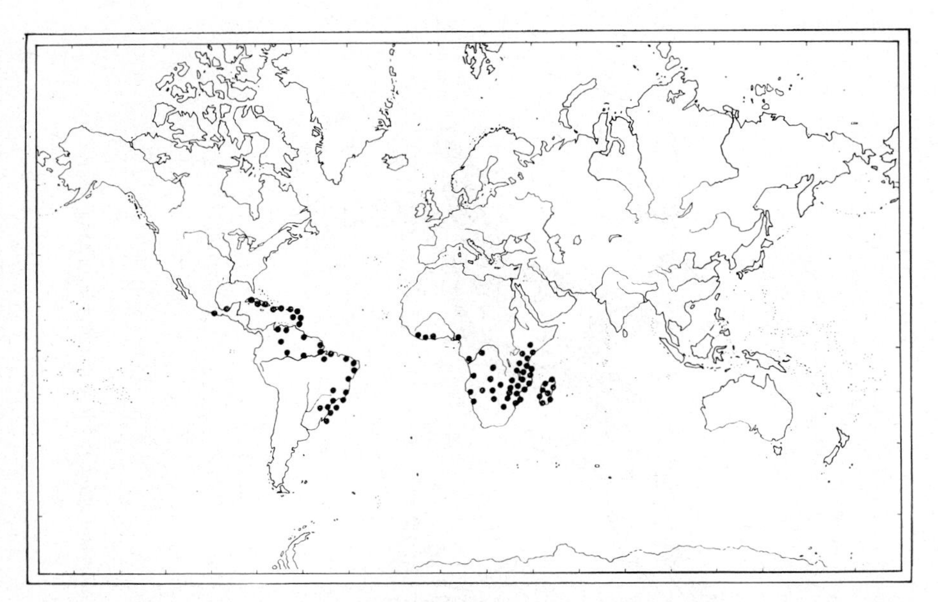

Fig. 24. Present-day distribution of *Hemidactylus mabuia* (after MÜLLER 1969). The species was originally restricted to Africa.

(*Erinaceus algirus*) has probably been introduced several times from North Africa to Malta (MALEC & STORCH 1972). *Erinaceus algirus* also occurs, probably having been introduced, on the Canaries (Teneriffe, Fuerteventura) together with introduced rats, house mice and rabbits (NIETHAMMER 1972). At the present day the only flightless mammals found on the Canaries are introduced forms but it is known from early Pleistocene fossils that the islands were formerly inhabited by autochthonous mammals e.g. *Canariomys bravoi* (CRUSAFONT-PAIRO & PETTER 1964).

Naturalisation of new species of animals in New Zealand has produced a basic transformation of the fauna. Examples of such naturalised species are the Australian brush-tailed opossum or kusu, the European red deer, the fallow deer, the wild boar, the common hare, the polecat, the stoat, the weasel and the hedgehog. Some of these species such as the red deer, are already being used economically. The brush-tailed opossum (*Trichesurus vulpecula*) lives in the tree crowns of the laurel and conifer forests, especially in trees of the genus *Metasideros*. By nibbling the crowns of the trees it is opening up the originally close canopy of the rain forests. The red deer cannot find suitable conditions in dense rain forest and therefore invade the forests affected by the brush-tailed possum. The result of the combined action of the possum and the red deer is the destruction of the forests.

The New Zealand flora also has many species introduced by man. Some

Fig. 25. The range resulting from introduction of the mosquito fish (*Gambusia affinis*); Close shading = original range.

examples are: common gorse (*Ulex europaeus*), tree lupin (*Lupinus arboreus*), dog rose (*Rosa canina*), blackberry (*Rubus fruticosus*), wild thyme (*Thymus serpyllum*) and foxglove (*Digitalis purpurea*). Compare in this connection the works of CLARK (1949), CUMBERLAND (1940, 1941, 1962), DIELS (1896), SCHMITHÜSEN (1968) and SCHWEINFURTH (1966).

The introduction of non-indigenous species of animals leads in most cases to considerable disturbance of the indigenous communities, or even to catastrophes. Damage of totally unforeseen extent can occur. This is shown for example by the history of the introduction of the musk rat into Central Europe (*Ondatra zibethicus zibethicus* cf. PIETSCH 1970), and of the Colorado beetle into the same area. Similar examples are the introduction of the rabbit into Australia (cf. RATCLIFFE 1959) or of the giant snail *Achatina fulica* into South-east Asia. Starting from East Africa this snail has penetrated in the last 200 years into the cultivated areas of Asia, of the Pacific islands including Hawaii, of California and of Florida. It is extremely injurious in the regions which it has colonised because of its high rate of reproduction, its great food consumption and because it acts as a vector for plant diseases. It had no effective enemies and therefore has been able to spread quickly through the plantations. Attempts are now being made to control it by the use of viruses, of poisoned bait and of its natural enemies among snails in the genea *Gonaxis* and *Euglandina*. The dates of introduction of *A. fulica* into the places that it has colonised are as follows: Madagascar 1761, Mauritius 1803, Reunion 1821, Seychelles 1840, Calcutta 1847, Mussoori in the foothills of the Himalayas 1848, Comoro Islands 1860, Ceylon 1900, Bombay 1910, Malayan Peninsula 1910, Singapore 1910, Riau Archipelago 1924, Borneo 1928, Amoy 1931, Java 1933, Formosa 1936, Hawaii 1936, Thailand 1937, Okinawa 1938, Bonin Islands 1938, Palau Islands 1938, Sumatra 1939, Caroline Islands 1939, Mariana Islands 1939, Marshall Islands 1939, New Ireland 1940, Hong Kong 1941, Manila 1943, New Guinea, 1943 New Britain 1943, California 1946, Florida 1966.

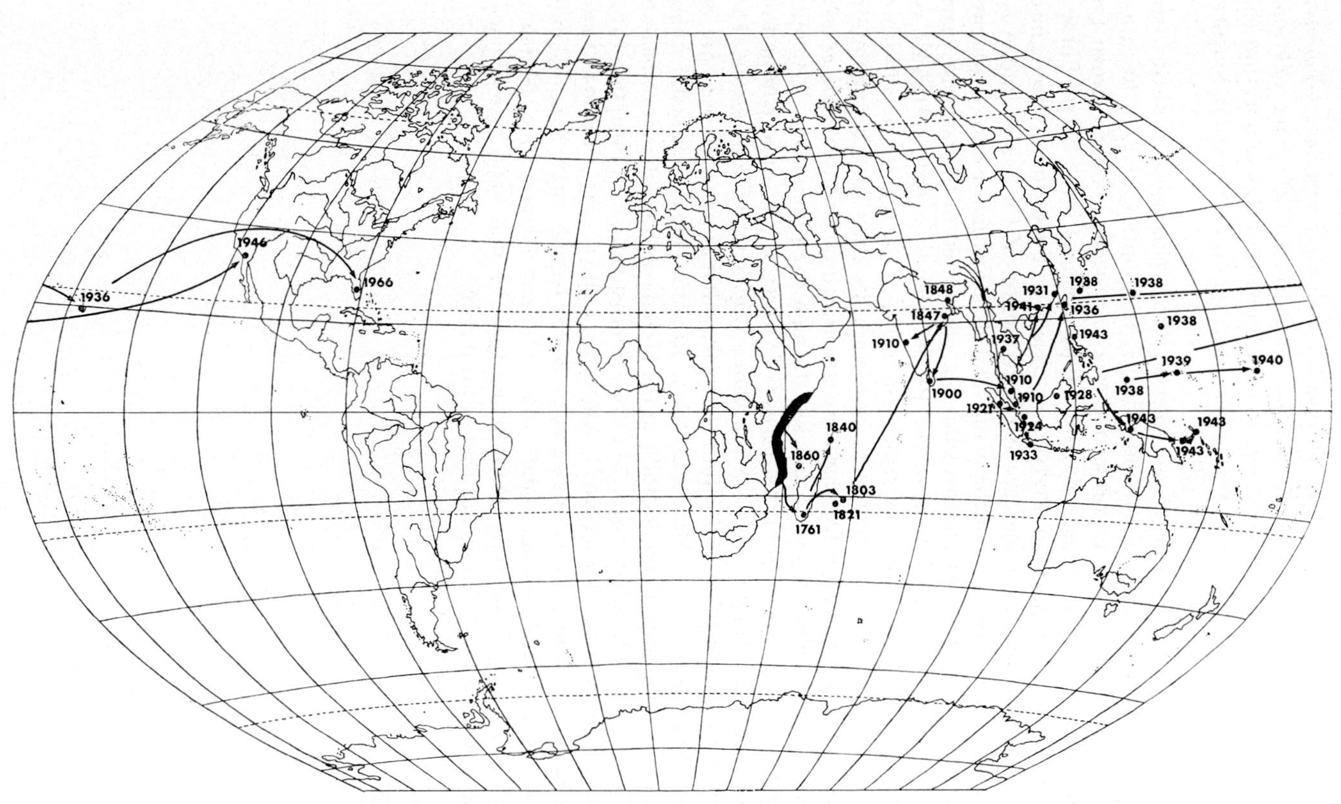

Fig. 26. Introductions of *Achatina fulica*; Black = range of *Achatina fulica hamillei.*

THE ZOOGEOGRAPHICAL REALMS

The gross arrangement of animal distribution in the biosphere led in the last century to the establishment of zoogeographical realms. Regardless of whether this gross arrangement is controlled by evolution or ecology, it holds to some extent for every group of animals. Difficulties only arise when attempting to set up realms that are equally important for all the different types of animals.

Proposals to divide the earth's surface into broad zoogeographical realms are essentially based on the geographical ranges of higher systematic categories such as genera and families. Particularly important are the works of SWAINSON (1935), SCHMARDA (1853), SCLATER (1858, 1874), GÜNTHER (1858), MURRAY (1866), WALLACE (1876), ALLEN (1878), REICHENOW (1888), WAGNER (1889), TROUESSART (1890), MÖBIUS (1891), MERRIAM (1892, 1894, 1898), BEDDARD (1895), LYDEKKER (1896), ORTMANN (1896), JACOBI (1900, 1939), KOBELT (1902), ARLDT (1907), HEILPRIN (1887, 1907), BARTHOLOMEW, CLARK & GRIMSHAW (1911), SHELFORD (1911), BRAUER (1914), MEISENHEIMER (1915), HESSE (1924), DAHL (1921, 1925), BOBINSKIJ (1927), MATRONNE & CUÉNOT (1927), BARTENEW (1932), PRENANT (1933), MARCUS (1933), EKMAN (1935, 1953), NEWBIGIN (1936), HEPTNER (1936), HESSE, ALLEE & SCHMIDT (1937), RENSCH (1931, 1950), BEAUFORT (1951), SCHMIDT (1954), ØKLAND (1955), LINDROTH (1956), SCHILDER (1956), DARLINGTON (1957), HUBBS (1958), GEORGE (1962), DE LATTIN (1967), NEWBIGJN (1968), UDVARDY (1969), NEILL (1969), and BANARESCU (1970, 1973). The concept of clearly definable realms, however, is constantly put in question by the great variety of species ranges and their varying histories. The present day ecological interpretation of lowland rain-forest, montana-forest, savanna and paramo, often within the smallest areas, produces such a confusion of types of range that a single subdivision and classification of the biosphere according to the regional concept becomes difficult. Despite these limitations, the zoogeographical realms are generally recognised and agreed units for groups that have been well enough studied, such as birds, mammals and reptiles. This is despite the fact that the limits between them are constantly put in question by individual groups of animals.

The proposals of SCLATER (1858) and WALLACE (1876) are still important today. Their splitting of the biosphere into three very broad realms, however, no longer seem sufficiently well grounded. These three realms were:

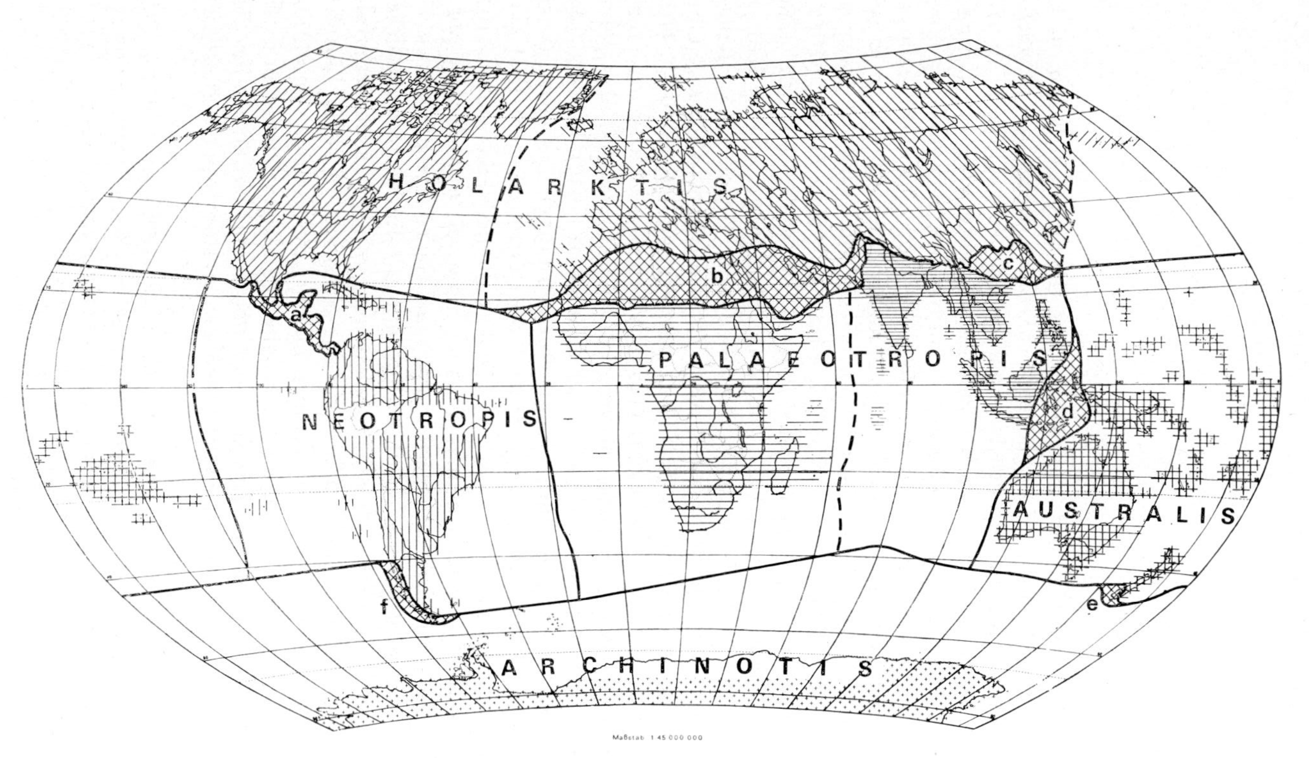

Fig. 27. Animal realms (after MÜLLER 1973); Crosshatched areas a, b, c, d, e, f, are transition zones.

Megagaea (Arctogaea)	=	North America, Eurasia, Africa, Arabian Peninsula, India and Indo-China.
Notogaea	=	Australia, Oceania and New Zealand.
Neogaea	=	South and Central America and the Antilles.

It is thus possible to set up a zoogeographical subdivision of the earth. This is based on comparisons of what is now known of the areas of distribution of vertebrates and of invertebrates, and takes into account the phylogenetic relations of the animals whose ranges are being considered. It essentially agrees with the phytogeographical realms, though showing striking peculiarities of its own.

The Zoogeographical Realms and Regions of the Earth

Realm	Region	Included Areas
1. Holarctic	a) Nearctic	North America; unlike the corresponding plant realm, it includes Florida and the Californian Peninsula, Greenland and the Mexican Plateau.
	b) Palaearctic	Eurasia (including Iceland, the Canary Islands, Korea, and Japan) and North Africa.
2. Palaeotropical	a) Aethiopian	Africa, south of the Sahara.
	b) Madagascan	Madagascar and its offshore islands.
	c) Oriental	India and Indo-China in the broad sense, as far as the Wallace Line.
3. Australian	a) Australian b) Oceanic c) New Zealand d) Hawaiian	Australia, New Guinea and the associated Islands, east of the Lydekker Line, part of New Zealand, Oceania and New Caledonia, Hawaiian Islands, Solomon Islands. In the present work the Solomon Islands, Middle and North New Zealand and Hawaii are left in the Australian Realm. These groups of islands however have so many peculiarities and phylogenetic connections with the Palaeotropical Realm that simply placing them in the Australian Realm is not really appropriate for all groups of animals.
4. Neotropical		South and Central America with the Antilles.
5. Archinotic		Antarctic, south-western South America and south-western New Zealand.

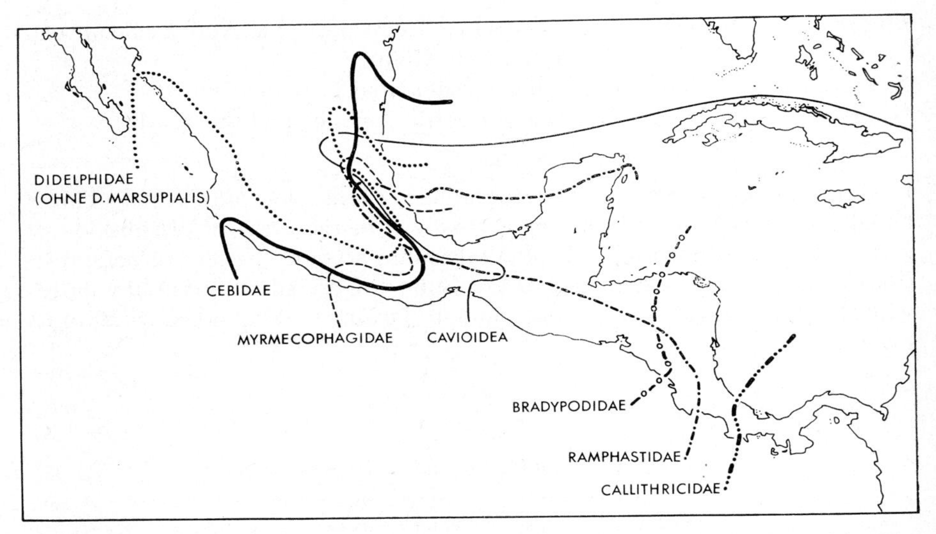

Fig. 28. The northern limits in Central America of families of South American origin. The limit of the Didelphidae does not apply to *Didelphis marsupialis*.

In general there are no sharp limits between the realms (ADAMS (1970) considers the problem of drawing boundaries). On the contrary, they meet each other at broad **zones of mixing and transition**. Those transitional zones in many cases have their own separate geological history, as in the example of Central America. In them, groups of animals which have long been indigenous are often overlaid by a 'stratum' of younger immigrants. For these reasons many transitional areas are seen by zoogeographers as independent zoogeographical regions. Clearly marked divisions between regions only exist where there are high mountains, broad arms of the sea, or ice deserts.

Wallace's Line – the famous line of division in the Indo-Australian transitional zone – has little significance for many groups of animals or for plants (WALLACE 1860). For other groups ,however, it is extremely well defined.

The transitional zone of Wallacea stretches from the Lesser Sunda Islands, Celebes and Lombok in the west to the Moluccas, the Kei and the Aru Islands in the east. Wallacea is separated from the Australian region by the Lydekker Line in the east (LYDEKKER 1896) and from the Palaeotropical Realm (Oriental Region) by Wallace's Line in the west (WALLACE 1876, MAYR 1944). Wallacea is essentially characterised by a mixed fauna of Australian and Oriental origin, but nevertheless has certain noteworthy endemics. These include the Celebes macaque (*Cynopithecus niger*) and the babirussa (*Babyrousa babirussa*) of Celebes which is a primitive representative of the Suidae.

Through the middle of Wallacea, between the Moluccas and Celebes, as well as between Timor and the Lesser Sunda Islands, runs the Weber Line. This was recognised by WEBER (1902) as a line where oriental and Australian groups of animals were about equal in abundance. WILKINSON (1969) and GRESSIT (1961) point out that it does not have the same significance for invertebrates as for vertebrates. SWELLENGREBEL and RODENWALDT have nevertheless shown that it controls the distribution of the vector mosquitoes of malaria i.e. the *Anopheles* species *A. aconitus*, *A. minimus*, *A. punctulatus*, *A. subpictus*, etc. It is worth noting that the Wallace Line was already established in its main features, by SALOMON MÜLLER in 1846. Müller's Line however placed Lombok and Sumbawa in the Oriental Region. Also MÜLLER considered, unlike WALLACE, that the line was determined ecologically. The eastern limit of the Australian marsupials largely agrees with the Wallace or Müller Line. In the south, however, marsupials are absent west of Wetar and Timor.

The boundary lines of Wallacea largely coincide with the 200 m depth contour of the sea, representing the depth to which the sea retreated during the Pleistocene. By contrast zoogeographical boundary lines on the continents are much less significant. This is true for example of the Reinig Line and the Johansen Line.

The Reinig Line follows the eastern bank of the River Lena and the River Aldan in Siberia and then passes over the Stanovoi and Yablonovyi mountains into Tienshan. It arose during the Ice Age and its causes are essentially historical.

The Johansen Line, on the other hand, is a contact region between Western and Eastern Palaearctic groups of animals (JOHANSEN 1955). It runs through the

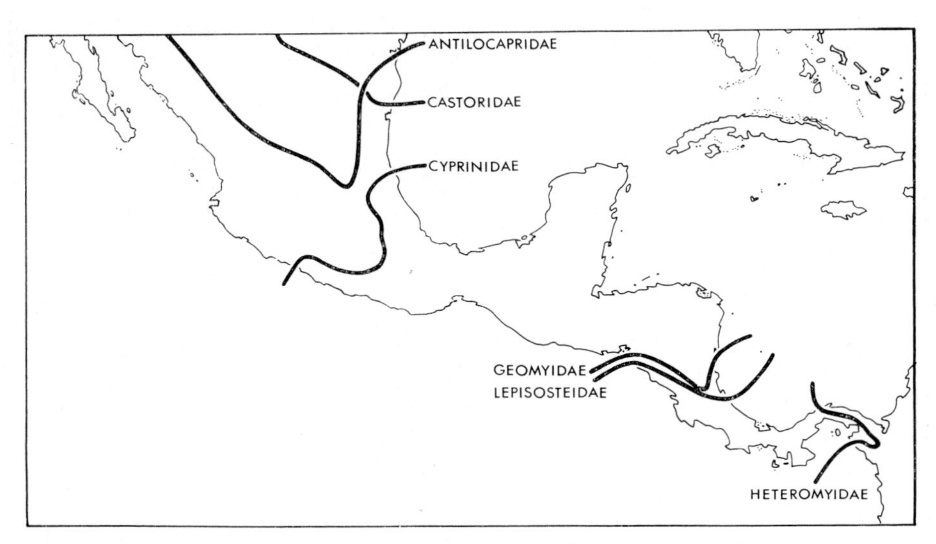

Fig. 29. The southern limits of North American families in Central America.

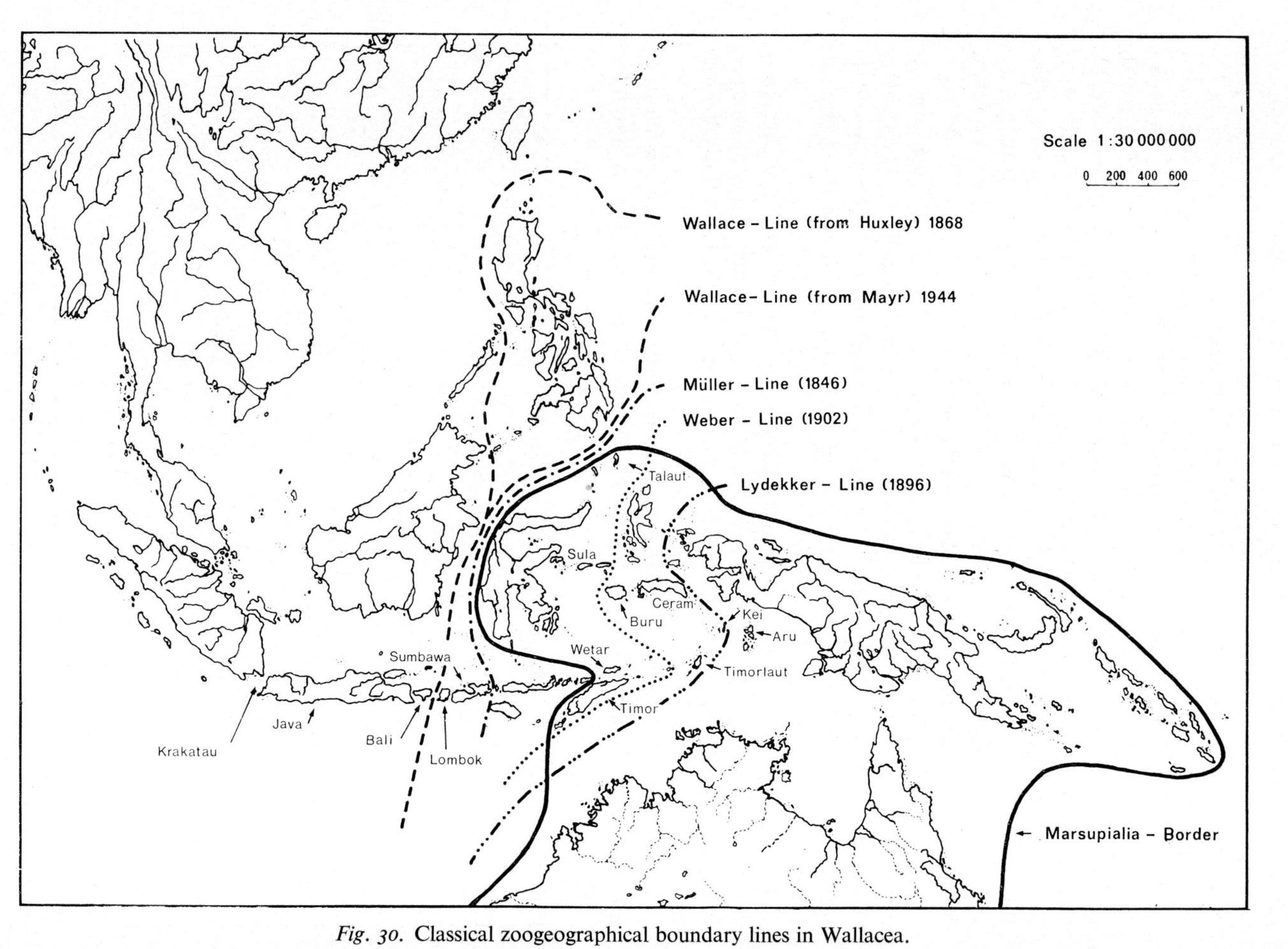

Fig. 30. Classical zoogeographical boundary lines in Wallacea.

Severnaya Zemlya Islands and the Taimir Peninsula and then follows the east bank of the Yenissie River – for which reason it is also called the Yenissei faunal boundary – as far as Altai. West of the Johansen Line lies the Central Siberian Plateau and this seldom allows western species, adapted to the lowlands, to extend any farther east. Contrariwise eastern, highland species meet their western limits at the western margin of the Central Siberian Plateau. Only when their ecological valency is very wide do they extend into the lowlands and thus pass west of the Johansen Line. Among eastern species which reach their western limit at the Johansen line are: the Siberian thrush (*Turdus sibiricus*), the eye-browed thrush (*Turdus obscurus*), the dusky thrush (*Turdus eunomus*), Naumann's thrush (*Turdus naumanni*), the Baikal teal (*Anas formosa*), and the falcated teal (*Anas falcata*). Among western species which reach their eastern limit at Johansen's Line are: the red-backed shrike (*Lanius collurio*), the sedge warbler (*Acrocephalus schoenobaenus*), the great snipe (*Capella media*), the spotted crake (*Porzana porzana*), and the velvet scoter (*Melanitta fusca*).

The limits between the Neotropical and Palaeotropical realms on the one hand, and the Holarctic realm on the other, have always been controversial and remain so. Central America is reckoned by most biogeographers to be in the Neotropical realm. Others, however, have made it a transitional zone between the Neotropical and Nearctic regions, though with predominantly South American groups of animals (MAYR, 1964, SIMPSON 1965, HERSHKOVITZ 1969, HOWELL 1969). Finally, some authors have raised it to be an independent realm, alongside the Neotropical realm (MERTENS 1952, KRAUS 1955, 1960, 1964, SAVAGE 1966). Part of the reason why individual workers have reached such different results is the varying powers of dispersal of the groups they have worked on (MÜLLER 1972, 1973). For groups with strong powers of dispersal, like mammals and birds, the peculiarities of the Central American region tend to be unimportant. Such groups have been worked on by SIMPSON (1940, 1950, 1965, 1966), MAYR (1964) and HOWELL (1969). For other groups, however, such as chilopods, diplopods amphibia, reptiles and gastropods, the singularities of the Central American region are very marked indeed (MÜLLER 1973).

Whatever views the different workers have come to concerning the placing of Central America, they all agree that in the lowland rain forest of that area the proportion of species of South American origin is astonishingly high. If the northern limits of South American families and the southern limits of North American families are plotted on a map of Central American, two things emerge: 1. There is a concentration of the northern limits of South American families in Central America, 2. It is possible to recognise a special barrier zone of the northern limits of South American families in Central America. This barrier zone coincides with the northern limit of the Central American lowland rain forest and with the 1500 m contour of the Sierra Madre in Mexico. In the Andean area, Northern American species penetrate far into South America.

A number of characteristic species are limited to the Central American tran-

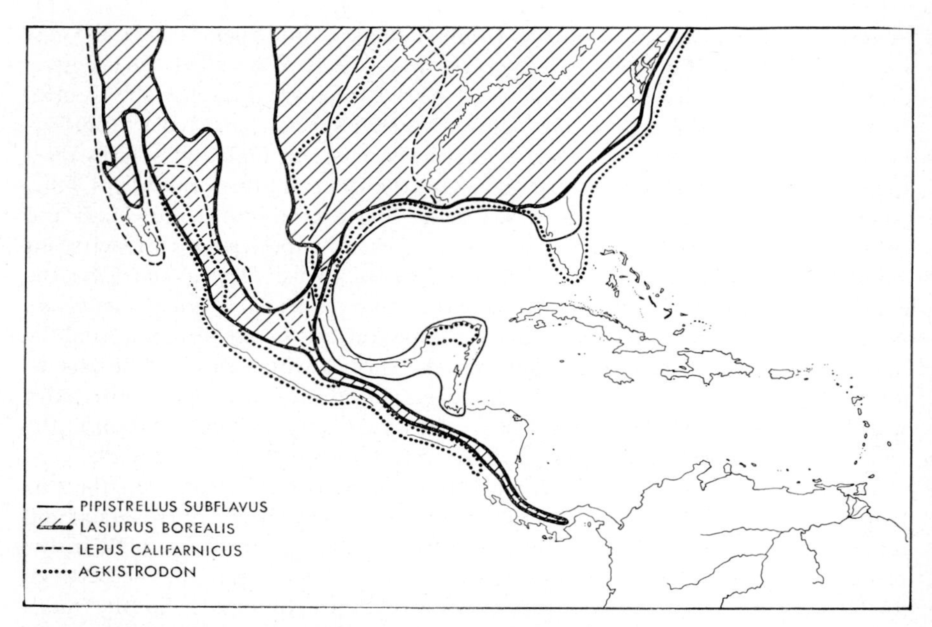

Fig. 31. The ranges of some Nearctic taxa in Central America.

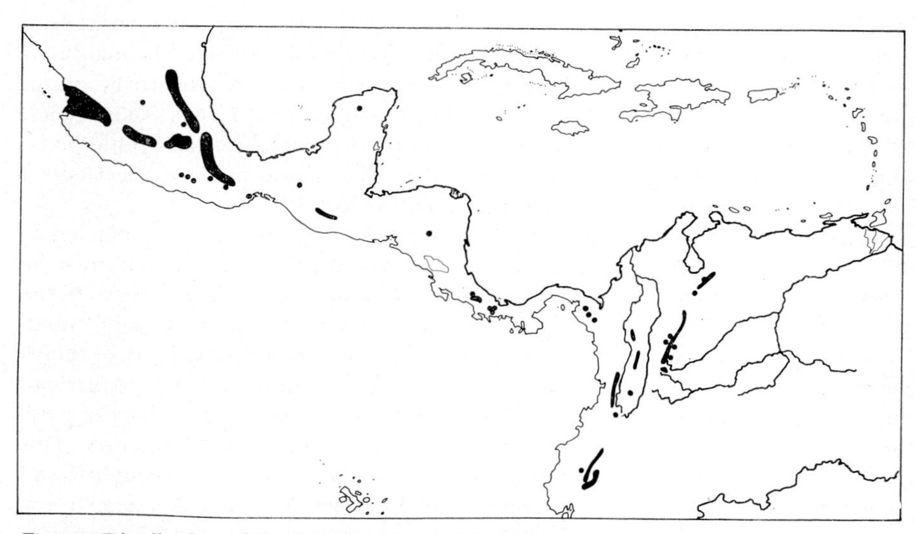

Fig. 32. Distribution of the insectivore genus *Cryptotis*. In Mexico and else where in Centra America species of this genus occur in the montane forests and paramos.

sitional region. These include the amphibia: *Eleutherodactylus alfredi, E. anzuetoi, E. bocourti, E. brocchi, E. decoratus, E. dorsoconcolor, E. dunni, E. greggi* and *Hyla robertmertensi*; the snakes: *Typhlops basimaculatus, Loxocnemus bicolor, Leptotyphlops phenos, Dipsas dimidiatus, Bothrops sphenophrys, B. yucatannicus*; the birds: *Campylorhynchus yucatanicus, Myiarchus yucatanensis, Agriocharis ocellata*; and the mammals: *Peromyscus yucatanicus* and *Sciurus yucatanensis*.

There is a similarly broad transitional region between the Palaeotropical realm (Aethiopian and Oriental regions) and the Holarctic realm. If the lines of separation proposed by a number of biogeographers were to be drawn on a map of north Africa and the Arabian Peninsula, the Sahara would be covered by a positive network of boundaries. The Sahara is not a uniform desert. In its driest central parts there are isolated blocks of mountains, such as Tibesti, Hoggar and Air. In these, Holarctic species spread far to the south while Aethiopian species reach northwards.

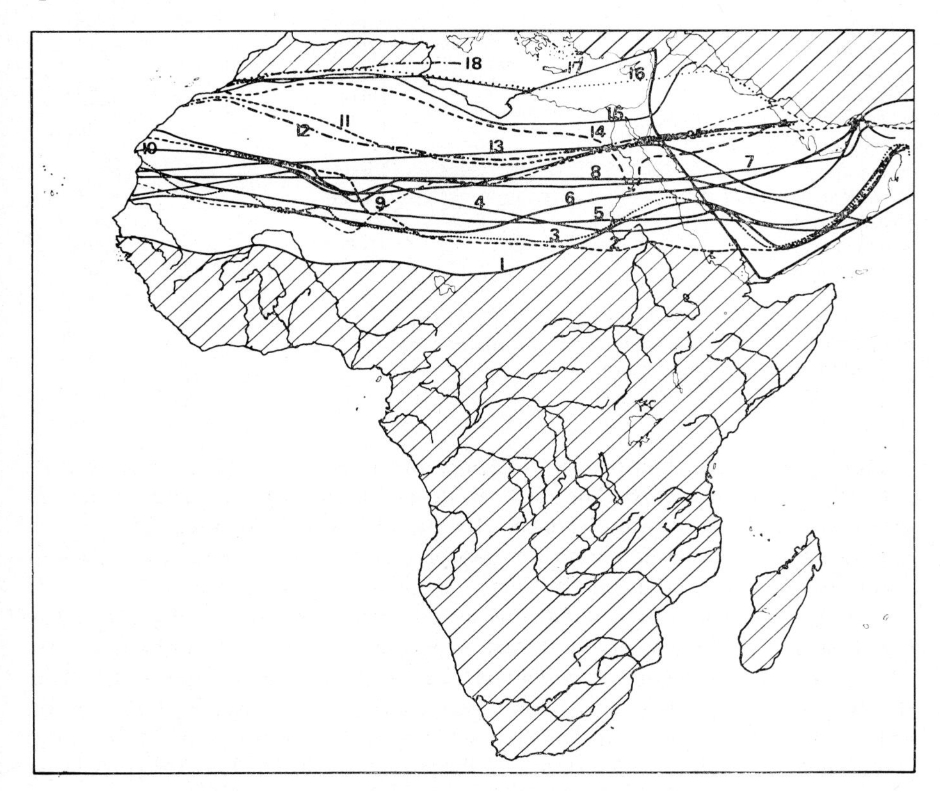

Fig. 33. The boundaries between the Palaearctic and Aethiopian realms as proposed by various authors.

Table II: Species of tree in Tibesti (after SCHOLZ 1967)

Species	Highest altitude of occurrence	Distribution of species
Hyphene thebaica	1200m	T
Ficus salicifolia (incl. *F. teloukat*)	2500m	T-SS
F. ingens	2000m	T-SS
F. gnaphalocarpa	1900m	T
F. sycomorus	1600m	EA
Boscia salicifolia	2000m	T
B. senegalensis	?1500m	T
Maerua crassifolia	1900m	T-SS
Capparis decidua	2000m	T
Acacia laeta	1800m	EA
A. nilotica (incl. *A. adstringens*)	1700m	T
A. stenocarpa	2300m	T
A. seyal	2100m	T-SS
A. raddiana	2200m	
A. albida	2200m	T
Balanites aegyptiaca	1800m	T-SS
Salvadora persica	1500m	T-SS
Tamarix aphylla (*T. orientalis*)	1500m	SS
T. gallica ssp. *nilotica*	2200m	SS
Erica arborea	3000m	EA, M
Calotropis procera	2200m	T-SS
Leptadenia pyrotechnica	1700m	T-SS

T – Senegambia – Sudan floral region (Sahel zone)
SS – North Africa – Indian desert area (Saharo-Sind)
M – Mediterranean floral region
EA – East African floral region

During the Postglacial period the arid central parts of the Sahara were considerably wetter than now. The stages of development of Lake Chad during the Holocene can be used to indicate the climate. During the late Quaternary three main stages of lacustrine conditions can be distinguished. These are the Chad freshwater-sea stage, the Bahr-el-Ghazal stage and the Lake Chad stage.

The Chad freshwater-sea is indicated by a bench at an altitude of 340 m in the Chad basin. This bench surrounded an inland sea with an area of about 320000 km². Thus, during the maximum of the Chad Sea, the area of water in the Chad Basin was only one-quarter less than in the present day Caspian Sea (438000 km²). The maximum level of the Chad Sea was controlled by the overflow of the river Chari into the Benue system at Fianga (GROVE & PULLAN 1963). A large number of deltaic piles formed by rivers at the former strand-line indicate strong fluviatile sedimentation during the Chad Sea maximum. The researches of ERGENZINGER (1967) permit a precise temporal correlation of this maximum.

According to the C_{14} dates of FAURE (in MONOD 1963) the Chad Sea stage lasted from 22000 to 8500 years ago. The maximum level at Trou au Natron – a lake in central Tibesti – was reached 15000 years ago (FAURE in ERGENZINGER 1968). This indicates that in late Würm times a pluvial climate was dominant above 2000 m. At this time Tibesti rivers were able to carry sediment into the Chad Sea.

In the period between 8500 and 5000 years ago, there was a sharp fall in sea level resulting from an extremely arid climatic phase. In this period the lake level in the Bodelé depression fell more than 60 m. 5000 years ago there started a tropical humid phase (BUTZER 1958) which is characterised as the Bahr-el-Ghazal stage.

The Bahr-el-Ghazal is the dry valley which now links Lake Chad with the Bodelé depression. It was cut during the arid phase of 8500–5000 years ago. At the time of the Bahr-el-Ghazal stage it connected the two terminal lakes which then existed in the Chad basin. These were Lake Chad itself, which was then 7 m higher and twice as large as now, and Lake Bodelé with a shore-line at an altitude of 240 m. Both lakes were fed by the streams of the southern part of the Chad basin. Only the southernmost rivers of the Wadai cut down into the 340 m bench of the former Chad Sea and built a new delta in Lake Chad during the Bahr-el-Ghazal stage. The more northern rivers no longer reached the lake. This lake level was reached about 3000 years ago according to the C_{14} dates of SERVANT (in ERGENZINGER 1968). The Bahr-el-Ghazal stage lasted from 5000 to 2450 years ago according to BUTZER (1958).

About 2450 years ago a strong arid phase began; Lake Bodelé dried up and the level of Lake Chad fell sharply. It was at this time that the Sahara first reached its present condition. The present-day period is characterised by dunes formed during this dry phase and is known as the Lake Chad stage (MESSERLI 1972, MOREAU 1966). In its present form Lake Chad is to be seen as the modest remnant of the various Quaternary lakes that preceded it.

The true desert fauna of Africa consists of numerous species which also occur in the Indian dry areas. Among plants such 'Saharo-Sindian' species amount to 70% of the whole. The close relationship with India is a justification for putting the Oriental and Aethiopian regions together in a Palaeotropical realm. The resemblances between the Aethiopian and Oriental Regions are greater than the resemblances of either with the Holarctic Realm. Thus in both regions the following families are found: among mammals: Tragulidae, Rhinocerotidae, Elephantidae, Hyaenidae, Hystricidae, Manidae, Pongidae, Cercopithecidae and Lorisidae; among birds: Nectariniidae, Pycnonotidae, Pittidae, Indicatoridae, Bucerotidae, and Pterochidae; among reptiles: the Chameleontidae; and among amphibians, the Rhacophoridae. As against this, the number of bird families endemic to either region is small, being 4 out of 67 families, for the Aethiopian region and only one family, the Irenidae, for the Oriental region.

267 genera of birds occur both in the Aethiopian and Oriental regions. Of Aethiopian species of birds, 69 extend to India, and 63 to Europe (MOREAU 1966).

A transitional region about as broad as that in North Africa exists in the area of China between the Palaeotropical and Holarctic Realms. The original subtropical forests have been extensively destroyed by human influence. Species adapted to open landscapes have partly replaced the original forest fauna. The fauna of Formosa, which during the Ice Age was connected to the mainland, shows a strong mixing of Oriental and Palaearctic faunas.

In the Indian deserts the palaearctic element is relatively high.

Palaearctic elements in different mammalian orders of the Indian desert (after PRAKASH 1974)

Mammal orders	Species occuring in the Indian desert	Palaearctic species	Oriental species	Palaearctic %
Insectivora	3	3	–	100
Chiroptera	11	4	7	36
Primates	2	–	2	–
Pholidota	1	–	1	–
Carnivora	13	9	4	69
Artiodactyla	4	2	2	50
Lagomorpha	1	–	1	–
Rodentia	16	7	9	44
Total	51	25	26	49

Finally, at the southern tips of South America and New Zealand there are other areas transitional to the Archinotic realm – the 'Old South'. These transitional areas have recently been particularly well-established by BRUNDIN (1965, 1972) and ILLIES (1965). The vertebrates that now occur in these regions are closest related to more northern populations. But many older invertebrate and plant groups indicate a close relationship of the southern tips of the continents (cf. MOORE 1972, VAN STEENIS 1972). Many families of crustaceans that were already present in the early Tertiary, and also Plecoptera and Chironomids, justify placing these transitional areas in the Archinotic Realm. This latter was already demanded by biogeographers of the last century both as a zoogeographical realm and as a land bridge. Its existence has been confirmed by many well-based zoogeographical examples worked out by zoogeographers of the present century.

Groups of younger and older immigrants occur in the various realms as though superimposed on each other. This produces a confusing complexity whose origin is one of the most exciting chapters of zoogeography. Exchange and mixture still continue and are at present accelerating. Nevertheless each

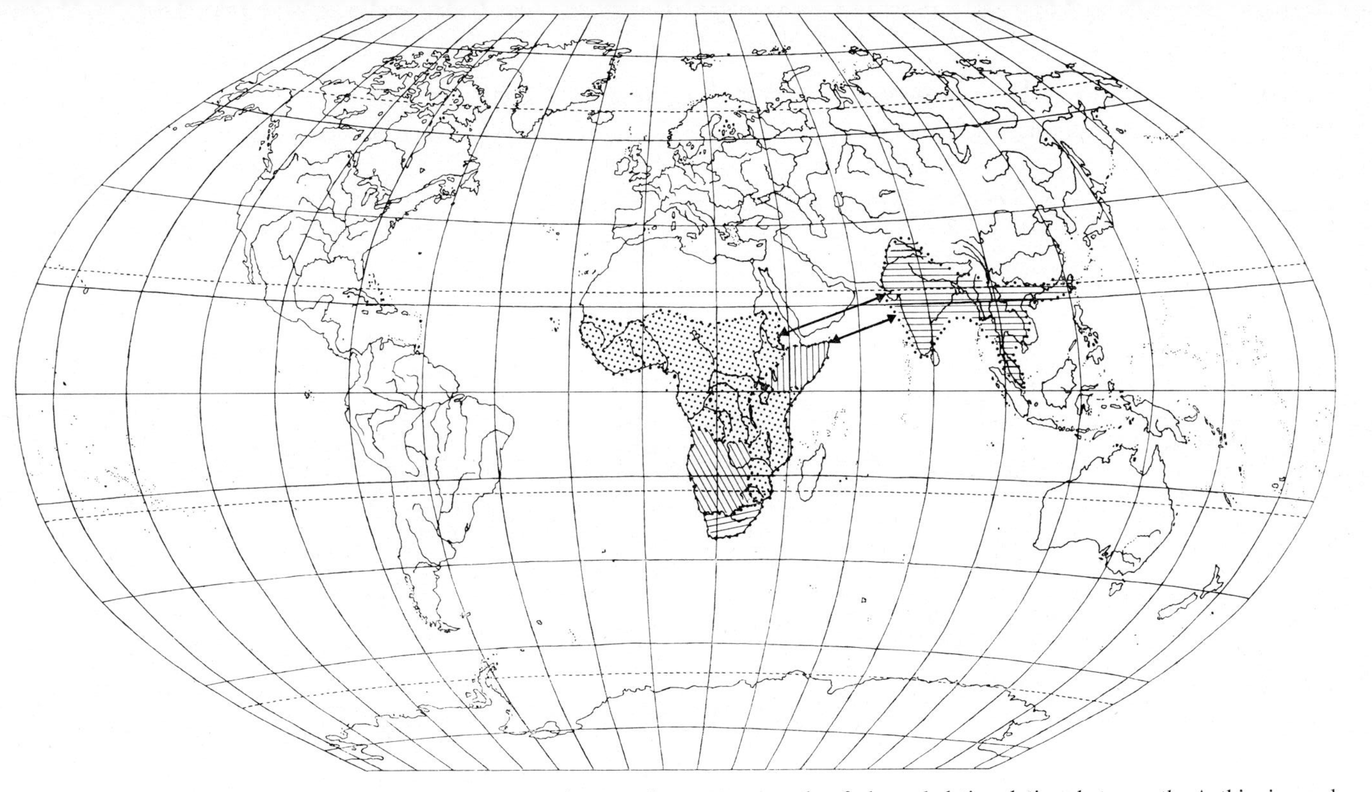

Fig. 34. The distribution of the *Prinia subflava* superspecies complex as an example of close phyletic relations between the Aethiopian and Oriental regions (after HALL & MOREAU 1970).

= *P. subflava*; = *P. somalica*; = *P. flavicans*; = *P. maculosa*; = *P. inornata*.

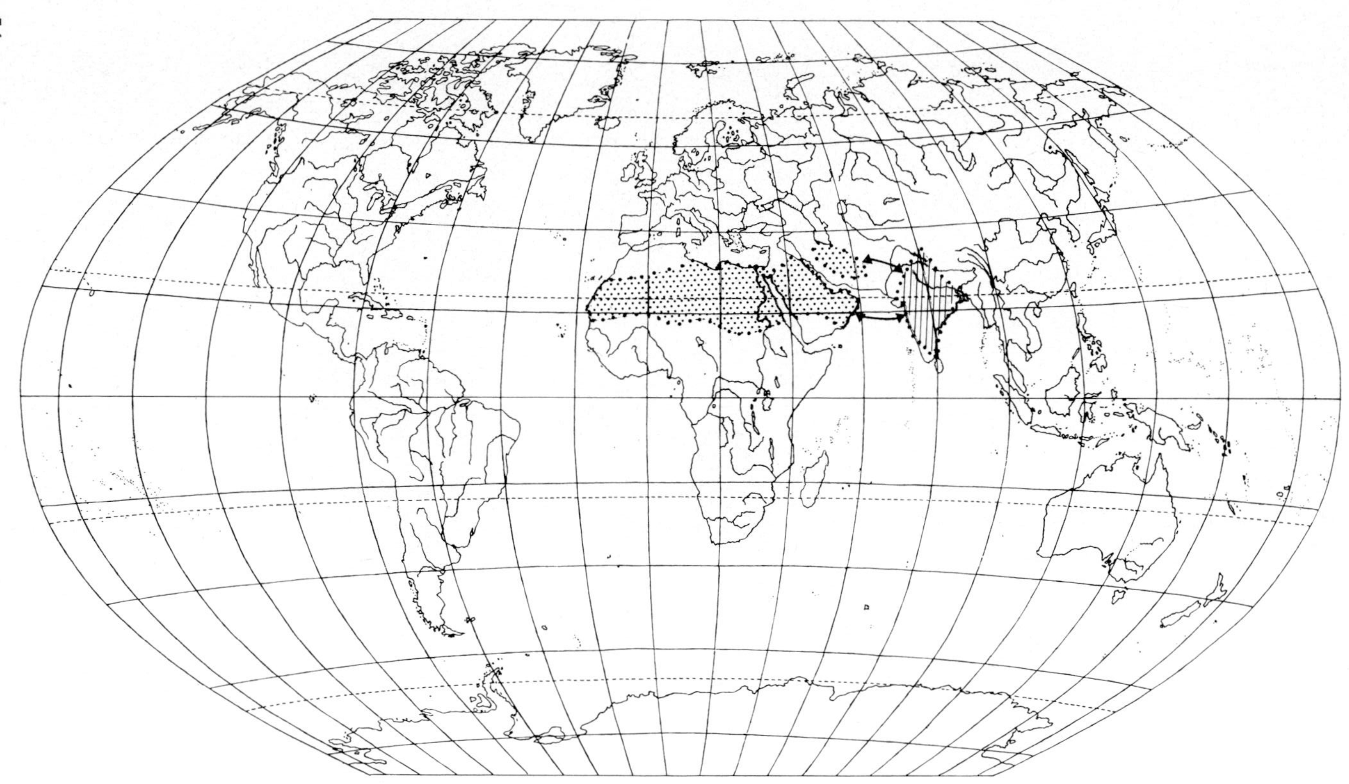

Fig. 35. Distribution of the superspecies *Ammomanes cincturus*. An example of the Saharo-Sindian distributional type (after HALL & MOREAU, 1970); [dotted] = *A. cincturus*; [vertical hatching] = *A. phoenicurus*.

realm still shows its own peculiar features that strike a scientific naturalist immediately he sets foot in one of them.

The best studied of all Realms is the **Holarctic Realm,** whose great zoogeographical importance was particularly well worked out by REINIG (1937). Characteristic groups of animals for the Holarctic realm are the moles (Talpidae), the genus Bison, the beavers (Castoridae), the Ochotonidae, the Zapodidae, the pikes (Esocidae), the fresh-water crayfishes (Astacidae), fishes of the subfamily Leuciscinae, the bumble-bee genus *Cullumanobombus,* butterflies of the genus *Colias* (*C. croceus* in the Palaearctic and *C. eurythemae* in the Nearctic) and *Choristoneura* (Tortricidae, on conifers). Also characteristic are the giant salamanders (Cryptobranchidae) occurring in East Asia and North America, the true salamanders (Salamandridae) and the olms or Proteidae. A monograph of the Holarctic Sesiidae (Lepidoptera) based on the phylogenetic and systematic methods of HENNIG (1957, 1960) shows the close relationship between North America and Eurasia (NAUMANN, 1969) (For the Bufonidae c.f. FLINDT and HEMMER 1972).

The causes of this close relationship are to be sought partly in the history of the two regions (c.f. DEEVEY 1949, KURTEN 1968) and partly in ecologically similar environments – continuous tundra and taiga zones etc. Numerous

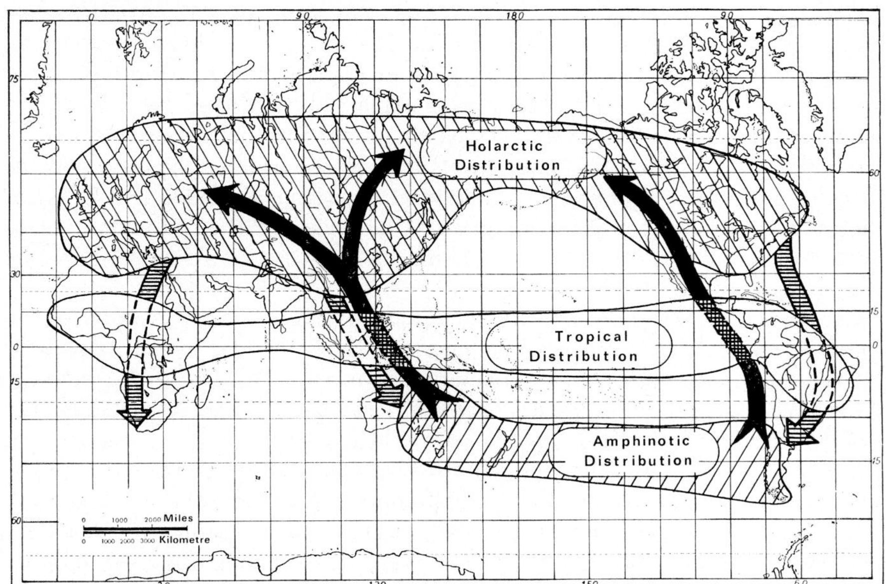

Fig. 36. Intercontinental connections in geographical range (after ILLIES 1970).

= Migration routes of southern groups (Siphlo nuridae).

= Migration routes of northern groups.

groups of plants have a Holarctic type of distribution. Examples are: Aceraceae, Betulaceae, Caryophyllaceae, Cruciferae, Fagaceae, Juglandaceae, Primulaceae, Ranunculaceae, Rosaceae, Salicaceae, Saxifragaceae and Umbelliferae. This is even fairly usual for species e.g. *Equisetum arvense* (common horsetail), *Cardamine pratensis* (lady's smock), *Zostera marina* (Common grass wrack or eel grass), *Angelica* (cf. PIMENOV 1968). The last land connection by way of the Bering Straits still existed in the Würm Glacial and allowed an exchange of the flora and fauna between the areas that are now separated (GRAHAM 1972, HOPKINS 1967, WOLFE & HOPKINS 1967, YURTSEV 1972).

Many cultivated plants in the arctic and boreal environments of North America and Europe have common sources (e.g. KONITZKY 1961). Species of the Eurasiatic tundra, such as the reindeer, of the taiga forest, such as the three-toed woodpecker, and of the deciduous forest, such as the red deer, are represented in America by closely related species or subspecies – a phenomenon known as vicariance. As ILLIES showed (1971), the faunistic resemblance of the northern continents is much greater than that of the southern continents. Despite these resemblances there are a number of endemics to the Palaearctic and Nearctic regions which indicate that the faunas of the northern continents are more distinct from each other than the floras are. Nevertheless it would not be justified to speak of the Palaearctic and Nearctic as realms but only as regions of the Holarctic realm. This was pointed out by HEILPRIN (1887), BLANFORD (1890), LYDEKKER (1896) and REINIG (1937). The close phylogenetic relationships of the Nearctic and Palaearctic bird fauna were explained by MAYR (1963) as due to immigration groups of different ages passing from Eurasia to North America. Among the oldest Eurasiatic immigrants into North America he reckoned the Gruidae, Strigidae, Columbidae, Cuculidae, Corvidae, Turdidae and Odontophorinae. Among the youngest immigrants he reckoned the Tytonidae, Alaudidae, Hirundinidae, Certhiidae (cf. THIELCKE 1965, 1969), Sylviidae, Laniidae and Fringillidae.

Groups of animals limited to the Nearctic region are, among mammals: the rodent families, Aplodontidae, Geomyidae and Heteromyidae (which, of course, like the Soricidae, also occur in northern South America), and the prongbuck (Antilocapridae). Among reptiles there are: the Californian Anniellidae; the Helodermatidae, which are adapted to the arid regions of the south western U.S.A. and Mexico, and the Gerrhonotinae which push far to the south in the mountains of Central America. Among amphibia, there are: the systematically isolated Ascaphidae, which formerly were placed with the 'primitive' New Zealand frogs of the family Leiopelmidae; the tailed batrachian of the families Ambystomidae (including the famous axolotl), Amphiumidae and Sirenidae. The North and Central American turkeys (Meleagrididae) were formerly considered as a Nearctic family of birds.

Like the Nearctic region, the Palaearctic region also has a number of characteristic groups of animals. These include the Siberian salamanders (Hynobiidae), disc-tongues, (Discoglossidae) which extend as far as the Philippines,the true slow-worms (Anguinae), the hedge sparrows (Prunellidae), the dormice (Glirinae), the Spalacidae (European mole rats), the rodent family Seleviniidae which is represented by only one species, the pandas, which are probably related to the bears, and the chamois (Rupicaprinae).

The **Neotropical realm** essentially includes South America, the Antilles and large parts of tropical Central America. Its flora and fauna is extraordinarily rich in species and in endemics. Endemic plant families include Marcgraviaceae, Nolanaceae, Bromeliaceae, Cactaceae, Tropaeolaceae and endemic palm genera such as *Jubaea* and *Mauritia*. Among animals endemic forms include: the marsupial families Didelphidae (though the opossum has been able to move into large parts of North America during historical times) and Caenolestidae; among placental mammals, the anteaters (Myrmecophagidae), the sloths (Bradypodidae), the Armadillos (Dasypodidae), the broad-nosed monkeys (Ceboidea), the rodent families, Caviidae, Hydrochoeridae, Dinomyidae, Dasyproctidae, Chinchillidae, Capromyidae, Octodontidae, Ctenomyidae, Abro-

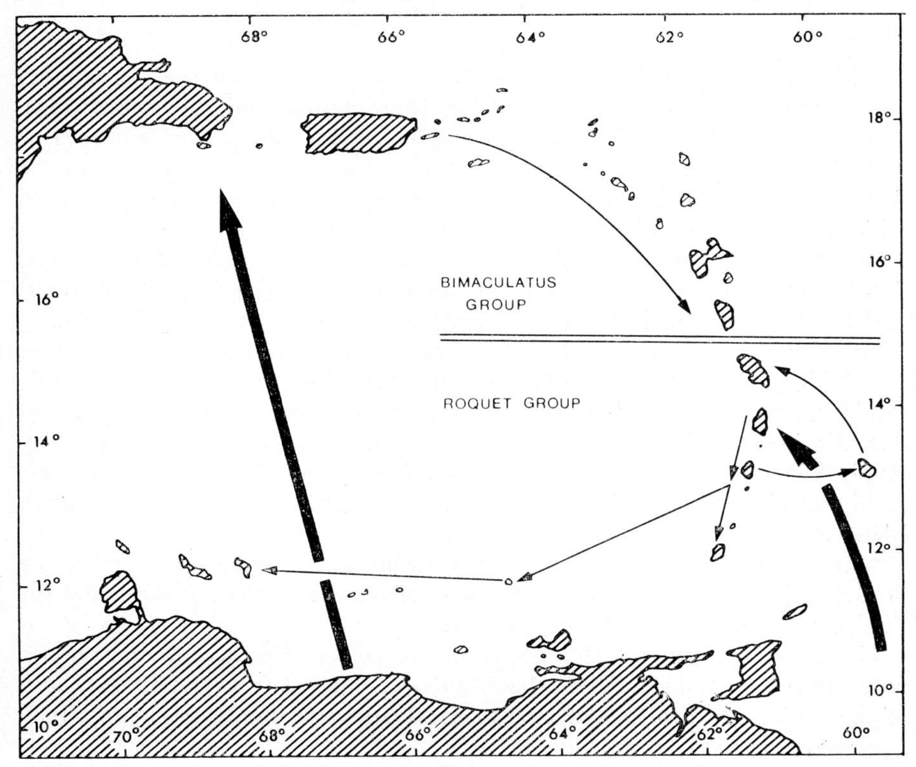

Fig. 37. The colonization of *Anolis bimaculatus* and *roquet* from South America to the Greater and Lesser Antilles (after GORMAN & ATKINS 1969).

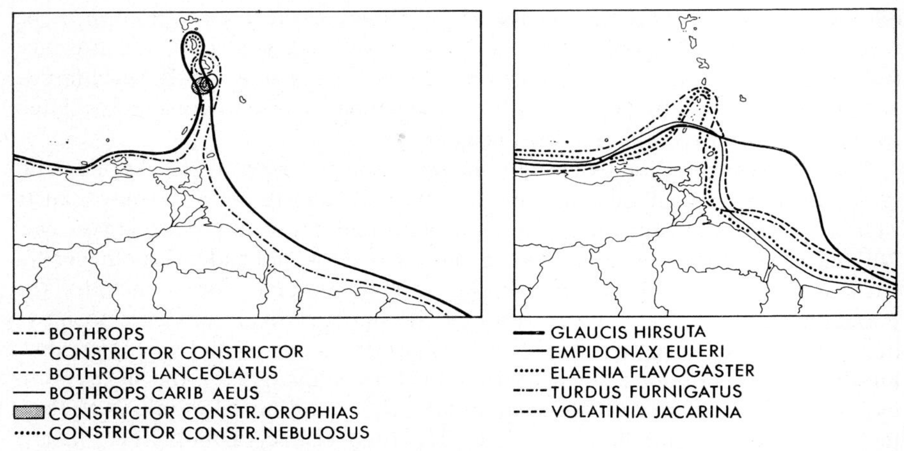

Fig. 38. The northern limits of South American taxa in the region of Trinidad and Tobago.

comidae and Echimyidae, the peccaries (Tayassuidae), and the bat families Desmodontidae, Natalidae, Furipteridae, Thyropteridae and Phyllostomatidae. The insectivore family of the Solenodontidae is restricted to the Antilles. It is virtually certain that the inventory of species of mammal in this realm is still far from complete. This is not true in the same sense for the birds although even with them new species have recently been described.

No less than 2926 species of birds with two endemic orders and 30 endemic families occur in South America. The endemic orders are the rheas (Rheiformes) and the tinamous (Tinamiformes). The endemic families are the: Rheidae, Tinamidae, Cochleariidae, Anhimidae, Cracidae, Opisthocomidae, Cariamidae, Psophiidae, Aramidae, Eurypygidae, Thinocoridae, Nyctibiidae, Steatornithidae, Bucconidae, Galbulidae, Rhamphastidae, Dendrocolaptidae, Furnariidae, Pipridae, Cotingidae, Formicariidae, Conopophagidae, Rhinocryptidae, Rupicolidae, Phytotomidae, Oxyruncidae, Coerebidae, Tersinidae, and Catamblyrhynchidae. The family Todidae is endemic to the West Indian islands.

Many South American families reach their northern limit at the Sierra Madre do Sul in Mexico. This holds for the marsupials except the opossum, and for the New World monkeys and the anteaters.

The humming birds, which are endemic to the New World, have 242 species in South America alone. They inhabit all environments from the highest parts of the Andes down to the Amazon lowland. It is striking, however, that they are absent in the Galapagos.

Particularly remarkable among the mammals are the marsupials, which otherwise occur only in the Australian realm. They have 87 species arranged in two genera and two families. Other striking forms are the endemic edentates (sloths, anteaters and armadillos), the broad-nosed monkeys, the tree porcupines, the chinchillas, the peccaries and the high proportion of endemic families of bats. The jaguar, the puma, the white-tailed deer and the rattle-snakes are forms whose ancestors only immigrated during the Ice Age, when, because of the North Andean uplift, the Panama Straits became land (SIMPSON 1950, MÜLLER 1973). During long periods of the Tertiary, South America was an island continent.

Among reptiles the predominance of snake-necked turtles (Chelidae) is remarkable. The leguans, which otherwise are only found on the Fiji and Tonga islands are also noteworthy as are the coral snakes (*Micrurus*), the fer-de-lance snakes (*Bothrops*, *Lachesis*), the giant snakes (*Boinae*) and the caimans. As many as 694 species of snake and 635 species of lizard are endemic to South and Central America (MÜLLER 1973).

Among amphibians, the tree frogs (Hylidae) and frogs of the families Leptodactylidae and Atelopodidae have the greatest number of species. The true frogs (Ranidae) are only represented in South America by one species and the lungless salamanders by twelve species; one of the latter inhabits Amazonia while the others are all from Colombia. The catfishes are represented by the endemic armoured catfishes (Callichthyidae) and by armoured catfishes of the family Loricariidae.

Some of the 2700 South American species of fishes, and their parasites, show remarkable African affinities. Examples are the Characidae, Cichlidae, Osteoglossidae (with *Arapaima gigas*) and the South American lung fish *Lepidosiren paradoxa*. Among invertebrates there are close phylogenetic affinities with Africa and New Zealand. Examples are found among the millipedes, the Spirostreptidae, the ostracods, the arachnids, the Chironomids, the Plecoptera, the Onychophora and the molluscs.

During the Quaternary horizontal displacements of vegetation occurred in South America. The Amazon forest was broken up by corridors of savanna during arid phases (HAFFER 1969, MÜLLER 1968, 1971, 1972, 1973). A large part of the richness in species of the various South American environments is due to these displacements and the isolation of individual populations that resulted from them.

The **Australian realm**, like the Neotropical realm, is characterised by a fauna and flora very rich in endemics (cf. BODENHEIMER 1959, KEAST 1959, PRYOR 1959). The Australian zoogeographical realm includes, as well as the mainland of Australia and Tasmania, also the areas of New Guinea, New Zealand, New Caledonia, East Melanesia, Micronesia, East Polynesia and Hawaii. On the basis of the vertebrates it is certain that New Guinea should be placed together with Australia, as older authors proposed. This is not so true for the Fiji Islands, the Solomon Islands, Micronesia, East Polynesia and the Easter Islands, and Hawaii. Because of their isolated position these areas have developed very much on their own (c.f. FRANZ 1970, PUTHZ 1972). To some extent they are as much Palaeotropical as Australian (cf. HOLLOWAY & JARDINE 1968).

The close affinity between New Guinea and Australia is due to the fact that because of eustatic falls in sea level, the Torres Straits have several times become dry land. The last land connection between Australia and New Guinea was only broken between 6500 and 8000 years ago (WALKER 1972 etc.). Faunal exchange occurred several times between New Guinea and Australia during the Pleistocene. The mountains of New Guinea were glaciated during the Pleistocene (LÖFFLER 1970, 1972, PAIJMANS & LÖFFLER 1972). In the rain forests of New Guinea a large proportion of Paleotropical forest species is present among the invertebrates.

Australia and the offshore island within the 200 m line have long been considered remarkable for the vertebrate groups that occur in them. In addition to the monotremes and marsupials, and the dingo introduced by man (*Canis familiaris dingo*), there are both in New Guinea and Australia numerous endemic specis of higher, placental mammals. KEAST (1968) has given a numerical comparison of marsupials and placentals in Australia and New Guinea.

Australia

	Genera	Species	Proportion of fauna
Monotremata	3	5	1.3%
Marsupialia	60	145	39.6%
Placentalia	51	214	59.1%

New Guinea

	Genera	Species	Endemics
Monotremata	2	3	2
Marsupialia	24	50	33
Placentalia	45	122	93
Total fauna	71	175	128

The fish fauna also has noteworthy endemics (WHITLEY 1959). The obligatory fresh-water spawners are worth mentioning; they are: *Neoceratodus forsteri*; *Melanotaeniidae*, *Macquaria*, *Percalates*, *Plectroplites*, *Maccullochella* and *Gadopsis*.

Because of the number of species in its bird fauna the Australian realm or Notogaea, has also been called Ornithogaea. In the 17th century it received the name of 'Terra psittacorum' because of the large number of parakeets and parrots. Endemic Australian bird families include: the bower birds (Ptilonorhynchidae), the emus (Dromicidae) which are adapted to open landscapes, the cassowaries (Casuaridae), the lyre birds (Menuridae), the megapods (Mega-

Antarctica. According to JANETSCHEK (1967) the algal flora of Victoria Land consisted of 80 Cyanophyceae, 50 diatoms and 20 Chlorophyceae (cf. also HIRANO 1965). 32 endemic species of lichen exist in the coastal region. The southernmost occurrence of higher plants – *Colobanthus crassifolius* and *Deschampsia antarctica* – is at 68° 12′S. Among small invertebrates the most important forms are: Tardigrada such as *Macrobiotus, Hysibius scoticus* (DALENIUS 1965, GRESSIT 1965); Acarina such as *Stereotydeus mollis, Tydeus setsukoae* and Collembola such as *Gomphiocephalus hodgsoni, Antarcticinella monoculata, Anurophorus subpolaris, Neocryptopygus nivicolus*. The Antarctic climate is the reason for the poverty of the Antarctic terrestial biota (RUBIN 1965).

Vertebrates are able to exist because of food chains that start in the sea. This holds especially for the penguins (Spheniscidae). These are an ancient group of birds, known as for back as the Eocene, whose phyletic relationships are still very obscure. They include 17 species and are excellently adapted to life on the Antarctic coasts by having plumage impermeable to water, a two- to three-centimetre thick layer of fat and various behavioural peculiarities which raise the basic metabolic rate. Most Antarctic species will not tolerate sudden rises in temperature. The emperor penguin breeds only on the coast of the Antarctic mainland, up to 1400 km from the south pole (PREVOST & SAPIN-JALOUSTRE 1965).

Other characteristic Antarctic penguins are *Eudyptes chrysolophus, Spheniscus magellanicus* (which, however, occurs as an accidental on the Brazilian coast as far north as Bahia), *Pygoscelis antarctica, Aptenodytes patagonica, Megadyptes antipodes* and *Pygoscelis adeliae*. Higher non-passeriform birds are represented by *Diomedea exulans, Stercorarius skua, Pelecanoides magellani, Oceanites oceanus* and *Chionis alba*. The passeriforms are only represented by a pipit (*Anthus antarcticus* (VOOUS 1965)).

Antarctic mammals also feed themselves by food chains that start in the sea. This is true for the seals, *Ommatophoca rossi, Hydrurga leptonyx, Lobodon carcinophagus, Leptonychotes weddelli* – and naturally also for the whales (*Balaenoptera physalus, B. musculus, Orcinus orca*). An important basic food for the species of *Balaenoptera* is the krill shrimp (*Euphausia superba*). The Antarctic Sea can be divided into individual, well delimited regions by means of its fish fauna (ANDRIASHEV, 1965).

The South Pacific forests of southern or false beech (*Nothofagus*) in Chile and New Zealand still contain a fauna which may have immigrated across Antarctica with these forests, perhaps as long ago as the Tertiary, following an 'arc' of land (HARRINGTON 1965, MÜLLER & SCHMITHÜSEN 1970). The geological structure of East Antarctica is different from that of West Antarctica. The Pre-Cambrian rocks of East Antarctica are an older continental complex than West Antarctica. The relatively young fold mountains connected with the Andean system – the South Antilles arc – arose at various periods from the Jurassic until the early Tertiary.

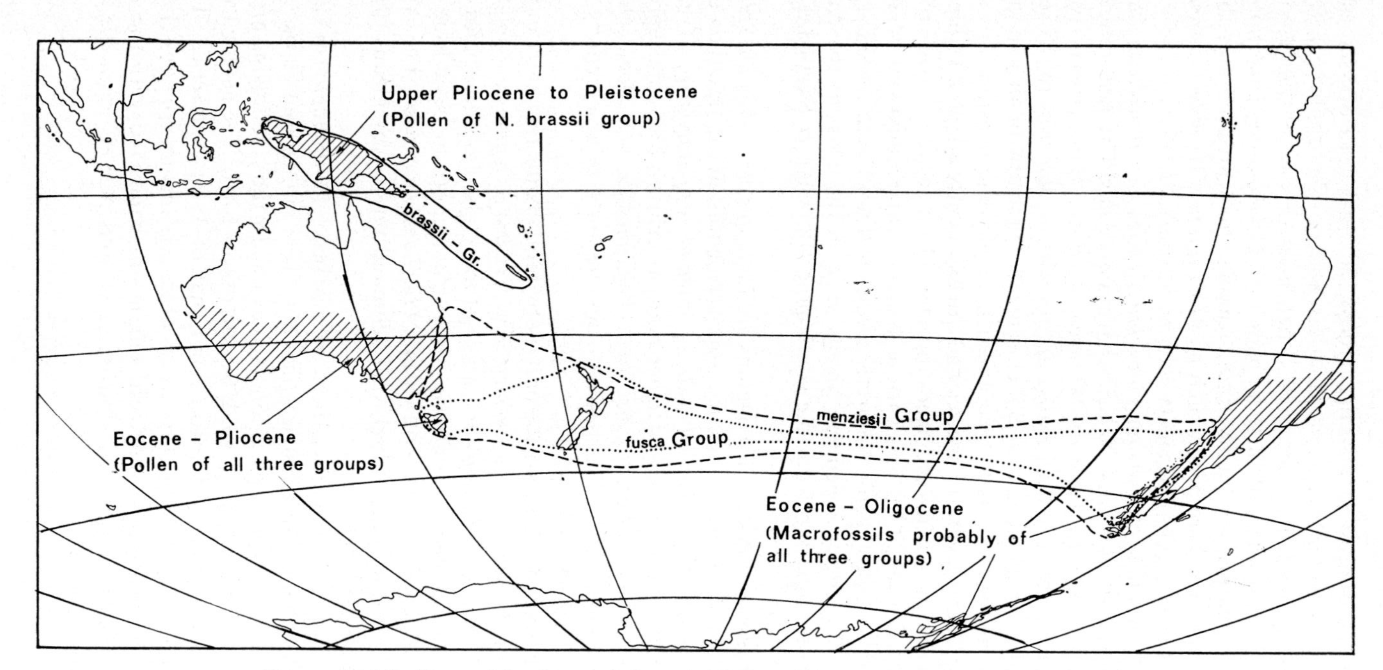

Fig. 39. Distribution and fossil records from *Nothofagus* (after MÜLLER & SCHMITHÜSEN 1970).

There are **zoogeographical realms in the sea** just as there are on land. The sea is divided into the deeps, the off-shore areas and the high seas and the oceanic realms are divided in the same way. The littoral fauna is confined to the shelf sea and therefore relatively subject to isolation effects. Following DE LATTIN (1967) it can be divided into three realms – (A) tropical (B) northern and (C) southern.

(A) **Tropical Realm**: 1. Indo-West Pacific Region, a) Malayan area, b) Southern Central Pacific area, c) Hawaiian area, d) South Japanese area, e) North Australian area, f) Indian area; 2. East Pacific region; 3. West Atlantic region; 4. East Atlantic region.

In the tropical realm coral reefs occur. These are a littoral biocoenosis of tropical seas. Their biotope depends in general on the lime-secreting activities of reef corals. Most coral reefs are essentially built up from the calcareous

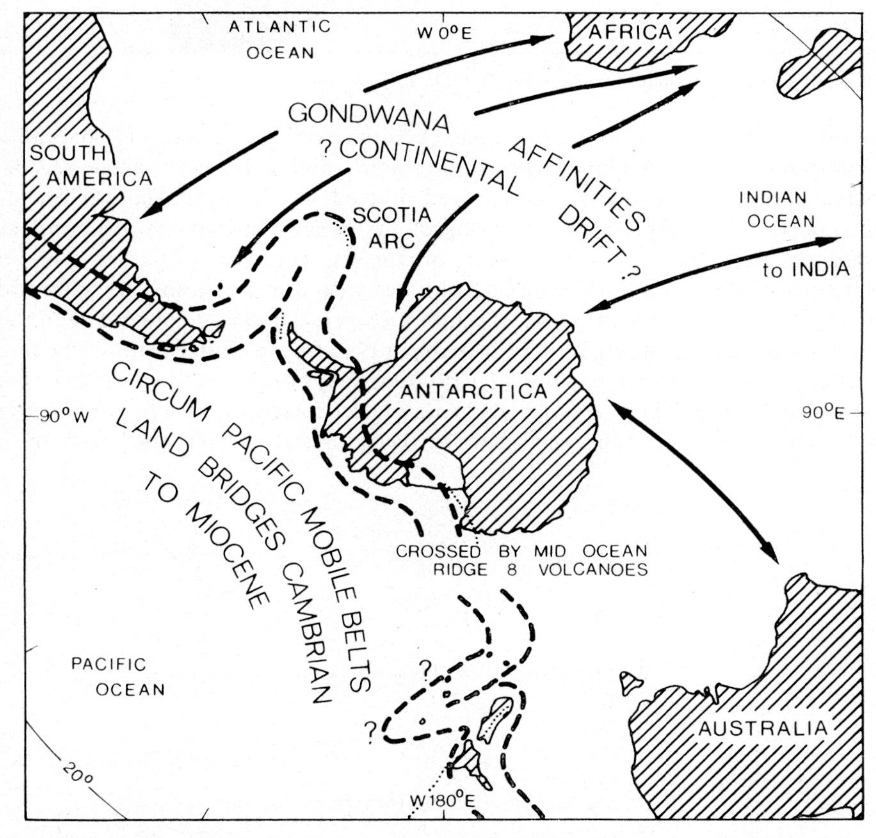

Fig. 40. Antarctica during the Cretaceous period (after HARRINGTON 1965). Fragments of Gondwana drift against the circum-Pacific arcs, which behave elastically are pushed inwards.

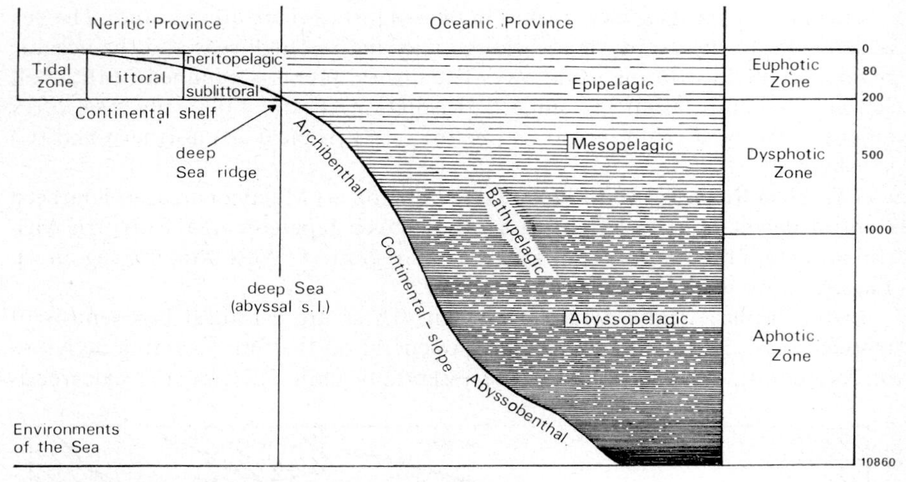

Fig. 41. The environments of the sea (after DE LATTIN 1967).

skeletons of Madreporaria (Poritidae, Acroporidae, Astralidae), Octocorallia (Tubiporidae) and Milleporidae; foraminifera and calcareous algae (*Lithothamnium*) are nevertheless also involved. Indeed, on the coral island of Funafuti, north of the Fiji Islands, a boring of 335 m showed that the foraminifera were the most important reef-building organisms, while the corals were only in third place. Reef building only occurs in seas with a mean annual temperature of 23.5°C or more. The presence of symbiotic, intracellular zooxanthellae in the tissues limits reef building to depths between 0 and 50 m. Furthermore, the fact that corals are restricted to salt water produces gaps in the reefs wherever rivers flow into the sea. There are various types of reef formation; there are fringing reefs immediately at the coast, barrier reefs separated from the coast by a

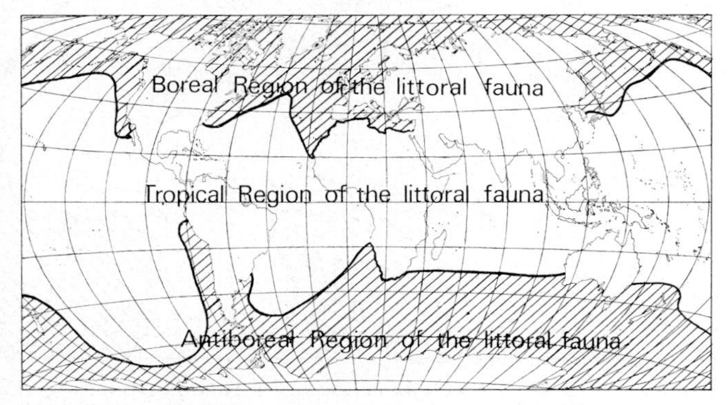

Fig. 42. Regional division of the littoral fauna.

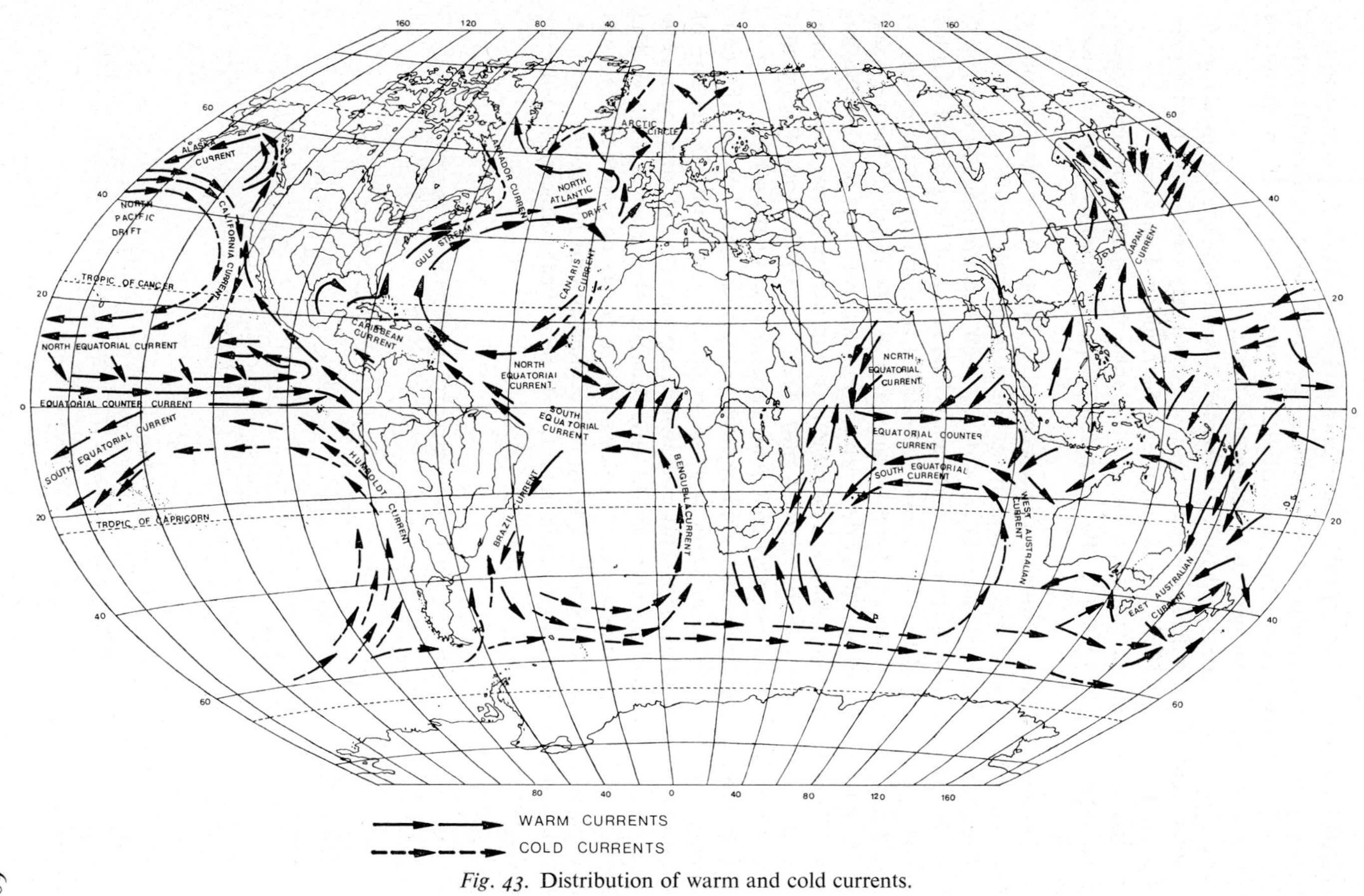

Fig. 43. Distribution of warm and cold currents.

channel and atolls which form a ring round a lagoon. DARWIN (1876) was able to show that these various types could only be understood in historical terms. The different types have greatly contributed to our knowledge of oscillation in sea level during glacial periods (FAIRBRIDGE 1962).

The fauna of the coral reefs is rich in species as would be expected in a tropical environment. Among the fishes that occur, the coral fishes are worth mentioning (Chaetodontidae, Pomacentridae, Scaridae, Serranidae, Labridae).

(B) **Northern realm**: 1. Mediterranean – Atlantic region; 2. Sarmatic region; 3. Atlantic – boreal region; 4. Baltic region; 5. North Pacific region; 6. Arctic region.

(C) **Southern realm**: 1. South African region; 2. South Australian region; 3. Peruvian region; 4. Kerguelen region; 5. South American region.

The groups of animals associated with the shelf seas are isolated from each other by wide oceanic barriers or continents. The fauna of the pelagial adapted to the open water of the high seas is controlled to a much greater extent by broad environmental factors, such as temperature. The fauna of the deep sea, or abyssal fauna, can be subdivided into regions controlled by the distribution of deep-sea basins and trenches. The same is true of the trench, or hadal, fauna. The deep-sea fauna can be divided into Arctic, Atlantic, Indo-Pacific and Antarctic regions controlled by particular geographical relationships.

The zoogeographical realms and regions just given, like those of the continents, are only the outer framework of a much more complex scheme of zoogeographical subdivision. This is shown by the fact that no one species has an ecological valency including all the environments that exist in a realm. The environment corresponds to the interplay of all the forces on which life depends in a particular place on earth. The aim of ecological and biogeographical research is to recognise this environment and its reciprocal connections with the organisms within it.

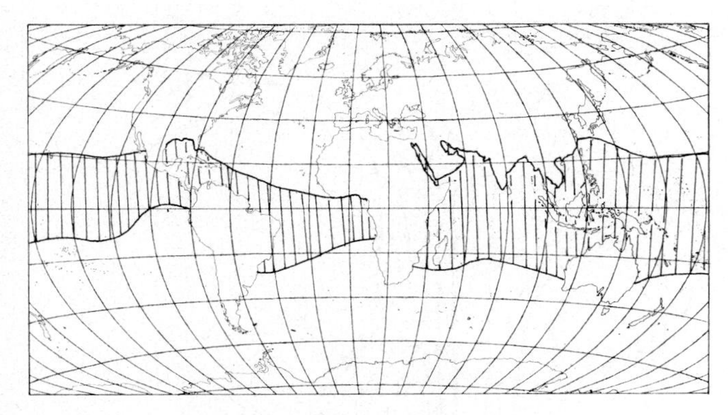

Fig. 44. Distribution of coral reefs (after DE LATTIN 1967).

Character gradients are correlated in direction with geographical and climatic factors. **Zoogeographical rules** reflect the connection between geographical and climatic factors and the differentiation of taxa. They express regularities in geographical variation with respect to climatic gradients.

1. BERGMANN's rule states that the body size of warm-blooded animals increases in colder climates.

2. ALLEN's rule states that the length of various appendages such as limbs, tail and external ears, decreases in colder climates.

3. RENSCH's hair rule states that hair is reduced in mammals with increase of temperature.

4. GLOGER's rule states that subspecies in warmer and damper areas are more strongly pigmented than in cooler and drier ones.

5. HESSE's rule states that the relative weight of the heart increases in cooler climates.

All these rules however only hold for homoiothermic organisms as was already noted by MELL (1929).

Animals cannot live for themselves alone. They depend on the plant cover, on other animals, on man and on non-living factors. All these produce the environment or biotope of the animal only by acting together. Species of animals which live in the same biotope and are able to reproduce and are dependent on each other form a life community or **biocoenosis** (MÖBIUS 1877). Biocoenosis and biotope form a mutually dependent unity which has its own dynamics. Individuals belonging to this unity occupy a particular position in it (equivalence of ecological position). Species diversity in a biocoenosis depends on the multiplicity of living conditions of the biotope (MACARTHUR & CONNELL 1970). The niche diversity of a biotope is critical when numerous species compete with each other in a biocoenosis (SHUGART & PATTEN 1972, NEVO et al 1972, CODY 1973, HAVEN 1973). Pressure of competition and the number of niches in a biotope control the **species diversity** (WILSON & BOSSERT 1973). The more uniform the environmental conditions, the less can competing species continue to live beside each other in the long term. The more diverse the environment is, or the quicker the environmental conditions change, so much the more readily can species co-exist.

The correctness of this rule can be demonstrated by simple experiment, such as carried out by GAUZE (1934), CROMBIE (1946) and FRANK (1952, 1957). CROMBIE put beetles of the genera *Tribolium* and *Oryzaephilus* together in glass vessels. If these vessels were kept sterile, without little tubes into which the smaller species *Oryzaephilus* could creep for protection, then the *Oryzaephilus* were wiped out after about 170 days. If little glass tubes were put in, both species could co-exist.

THE BIOMES

Life-communities fit into the major climatic and vegetational zones of the earth. Vegetational formations in the sense of GRISEBACH (1866) represent the basic framework of the life-communities. By vegetational formations are understood, for example, the tropical rain forests, the montane forests, the savannas, tundra, taiga and deserts. 'The basis for the maintenance of all life in the biosphere is the primary production of the green plants. In this the energy is stored which in photosynthesis has been converted from the light-energy of the sun's radiation to chemical energy. The latter serves to maintain all other living processes including those of man.'[1] (WALTER 1971).

Every vegetational formation has a specific structure (SCHMITHÜSEN 1968, DANSEREAU 1968). Vegetational formations, together with the animals adapted to them, are known as biomes. A biome is a higher category of biocoenosis. Thus the arid savanna biome is made up of the vegetational formation 'arid savanna' and of the savanna fauna characteristic of arid savanna. The biome concept goes back to CLEMENTS & SHELFORD (1939) and CARPENTER (1939). By biome, these authors understood the 'plant matrix with the total number of included animals'.

The biomes are arranged corresponding to the climatic zones of the earth. The biome includes all the life-communities within its territory and all their stages of development. The evolutionary concept, which is so critical for the biosphere as a whole is also important for the biome. 'The ever-changing diversity of an area in time' (i.e. the succession of biocoenoses) 'is included within the biome concept' (SCHMIDT 1969).[2] All biocoenoses that tend in their development towards the same terminal state or climax belong to the same biome. The latter, because it is essentially conditioned by climate, can only be basically altered by a fundamental change of climate. Together with the abiotic elements of a region, the biome constitutes a **Macro-ecosystem** (TANSLEY 1935, SOUTHWICK 1972, ELLENBERG 1973).

The biome concept is bound up with vegetational formations but the concept

1 'Die Grundlage für die Erhaltung allen Lebens in der Biosphäre bildet die primäre Produktion der grünen Pflanze, in der die Energie gespeichert ist, die bei der Photosynthese aus der Lichtenergie der Sonnenstrahlung in chemische Energie umgewandelt wird. Letztere dient zur Aufrechterhaltung aller anderen Lebensvorgänge, auch der des Menschen'.

2 'diese sich ewig wandelnde Mannigfaltigkeit eines Raumes in der Zeit wird durch den Biombegriff charakterisiert'

of the ecosystem, on the other hand, can be considered quite separately from them. Thus there are ecosystems which are not based directly on living plants, but which draw extensively on food brought in from outside. This is true, for example, for many communities of cave animals and for the hadal fauna of the deep sea trenches. It is critical for ecology that in an ecosystem there is a circulation of matter and energy. A knowledge of the equivalence of ecological position of a species in a particular ecosystem and its significance for stability or lability is therefore very important. An ecosystem can be almost independent of its surroundings; such are the autochthonous ecosystems as, for example, a primeval forest or a coral reef. Alternatively an ecosystem may draw on an influx of matter and energy derived from other ecosystems, in which case it is an allochthonous ecosystem. An ecosystem is a system of cause and effect that is limited in space. It is made up of abiotic elements and of biotic elements including man, and it has the power of self regulation.

In primeval landscapes unaffected by man, the hierarchical arrangement and spatial position of terrestrial ecosystems is usually correlated with the total geospheric situation of the region. In the smallest physico-geographical units of a landscape – the physiotopes – (SCHMITHÜSEN 1967, SUKACEV 1960, STODDART 1972) a specific local biocoenosis usually develops (MÜLLER 1972). Together with the locality where it occurs this biocoenosis forms an ecosystem (ecosystem I) which represents the smallest sort of ecosystem, apart from any key-species ecosystems interpolated in it. Examples of such key-species ecosystems are beaver ponds, ant hills, termite mounds and the nebka with tamarisk plants. The next higher type of ecosystem (ecosystem II) develops on the basis of landscapes made up out of physiotopes. These fit themselves into the ecosystem of a region (ecosystem III). The highest type of ecosystem is the geosphere with its living shell the biosphere.

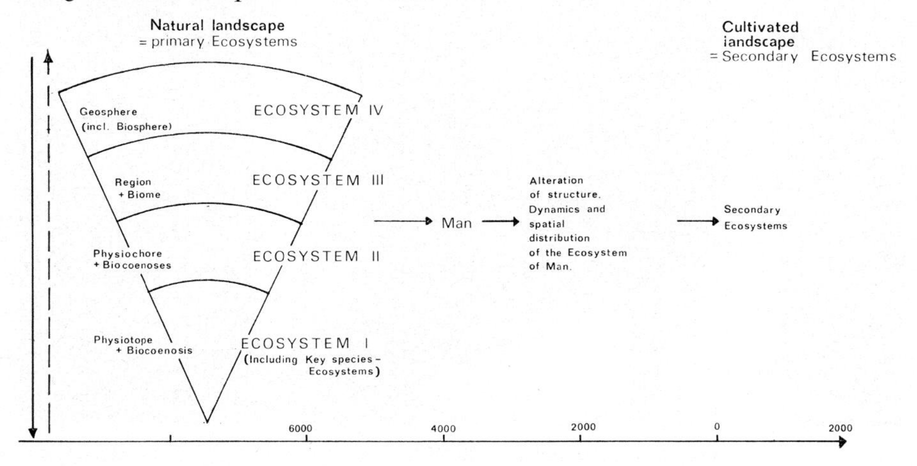

Fig. 45. The hierarchical subdivision of terrestrial ecosystems (after MÜLLER 1973).

The ecosystems of this hierarchy influence each other mutually. The primary ecosystems – those of the primeval landscapes – have been extensively and fundamentelly altered by man. The geosphere has been burdened all over the world with cumulative, summative and concentrative poisons so that the traditional division between natural and cultivated landscapes is largely no longer valid. At the present day only secondary ecosystems exist, though of course with a greater or lesser degree of human influence. Mankind is a new key-species which, since the Neolithic period, has altered the structure, dynamics and spatial distribution of the ecosystems. At the present time ecosystems do not necessarily agree at all with the physiotopes of a landscape. Their stability is essentially dependent on man. ELLENBERG (1973) had reviewed the 'nearly natural' ecosystems that exist on earth. But it is necessary to emphasise that other ecosystems do exist; these are 'human ecosystems' which must be considered exclusively from the human viewpoint. Even an industrial town is an ecosystem.

Many animals that live in particular biomes possess characteristic adaptations. These organic designs (**'Lebensformtypen'** of REMANE 1943, RUSTAMOV 1955) can be produced in different phyletic groups by convergence, and appear in spatially widely separated biomes. Thus soil animals in general possess char-

Fig. 46. A beaver lake in Parc Tremblant (Canada 1972); An example of a key species ecosystem.

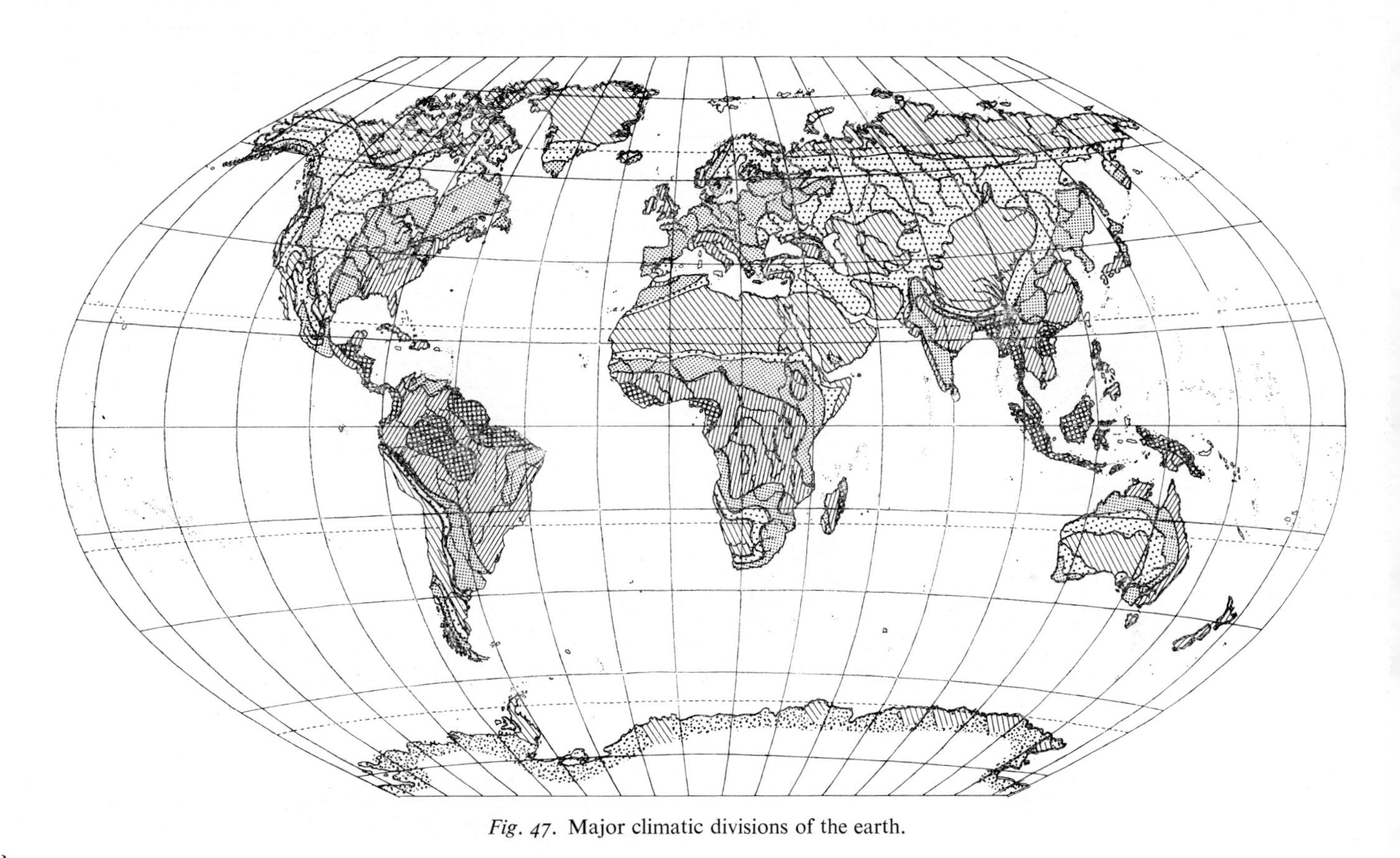

Fig. 47. Major climatic divisions of the earth.

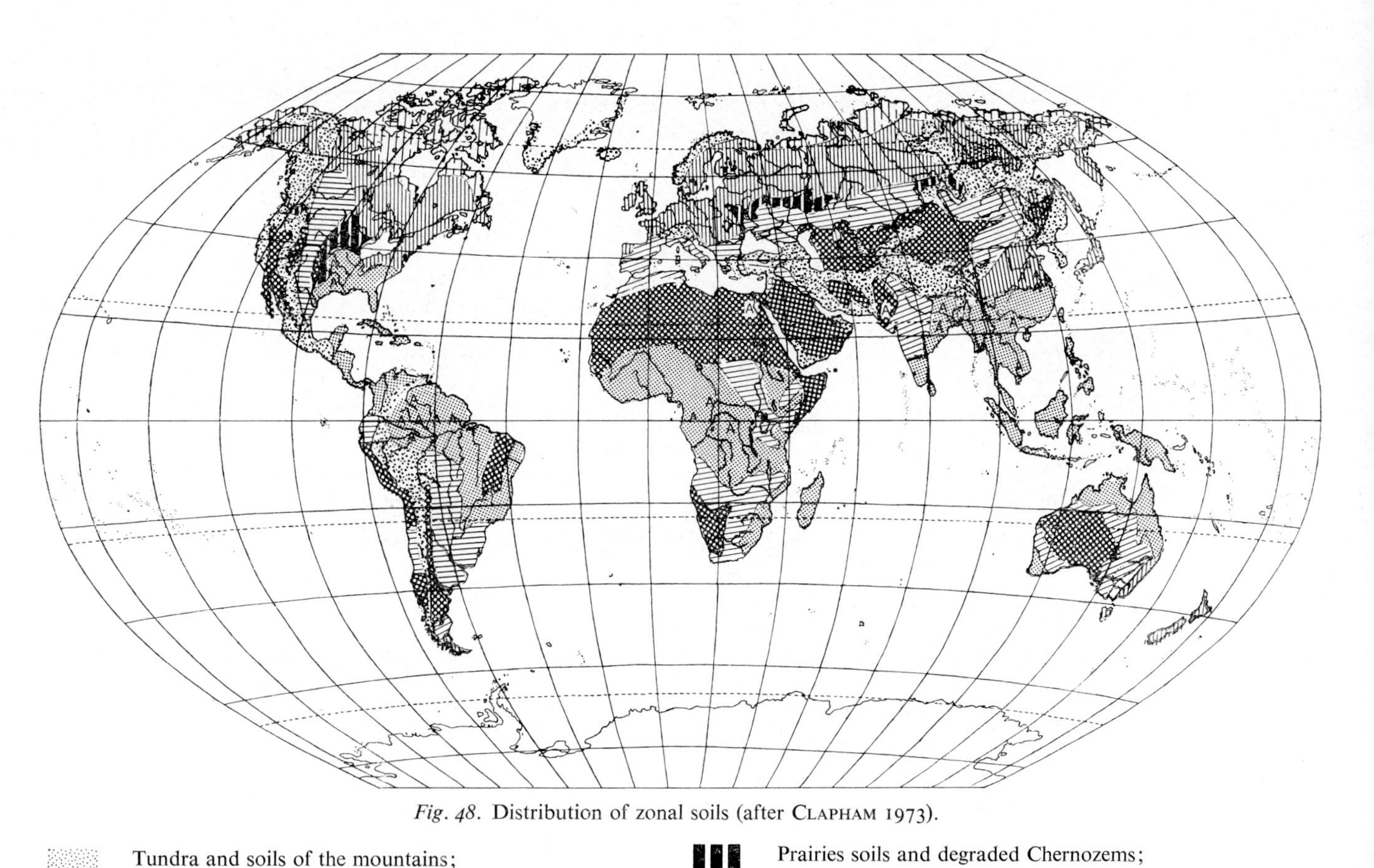

Fig. 48. Distribution of zonal soils (after CLAPHAM 1973).

- Tundra and soils of the mountains;
- Podzols and Gray-Brown, Podzolic soils (with Brown Forest soils etc.);
- Latosols;
- Prairies soils and degraded Chernozems;
- Desert and Red Desert soils;
- Chernozems, Chestnut, Brown and Reddisch-Brown soil

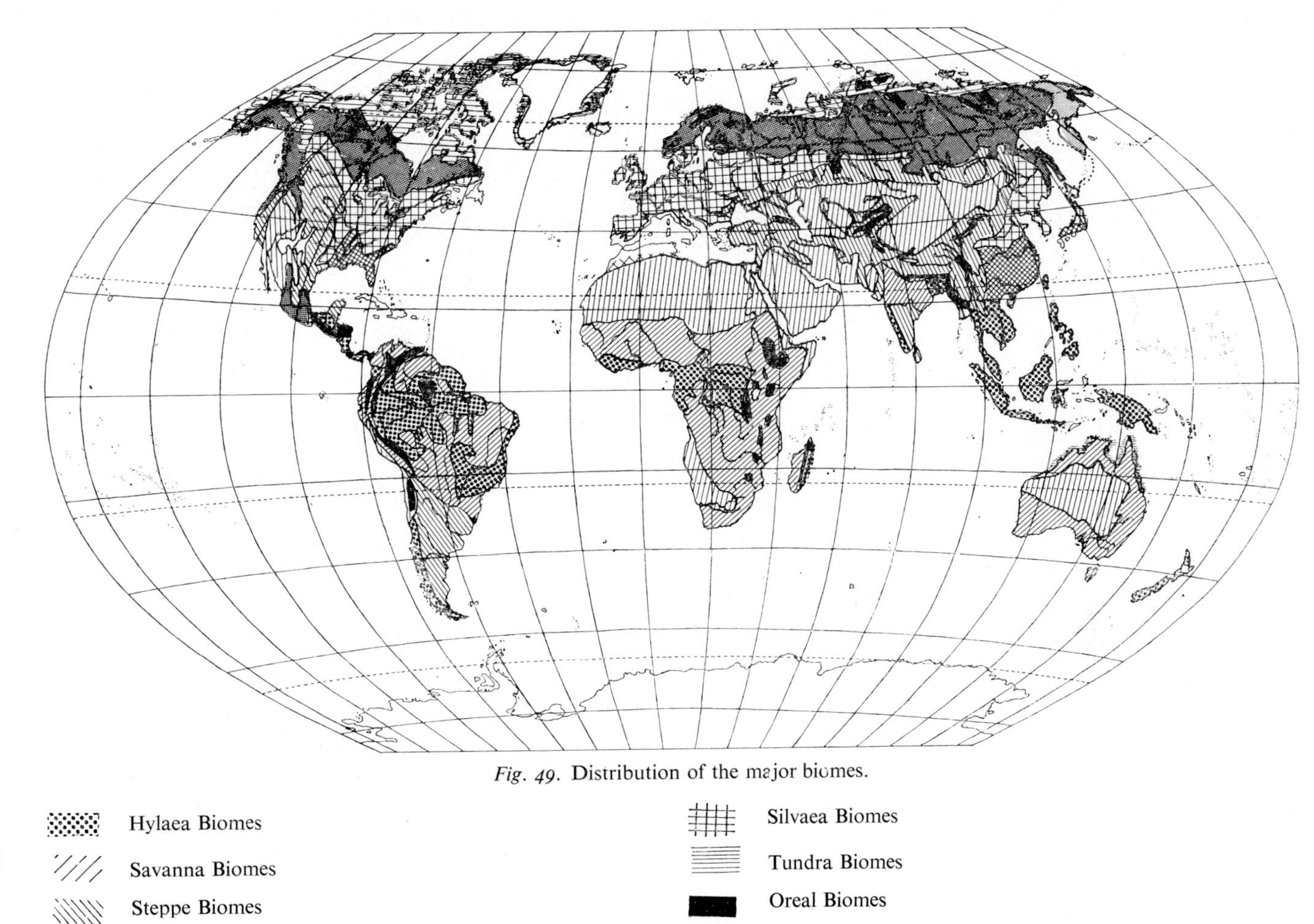

Fig. 49. Distribution of the major biomes.

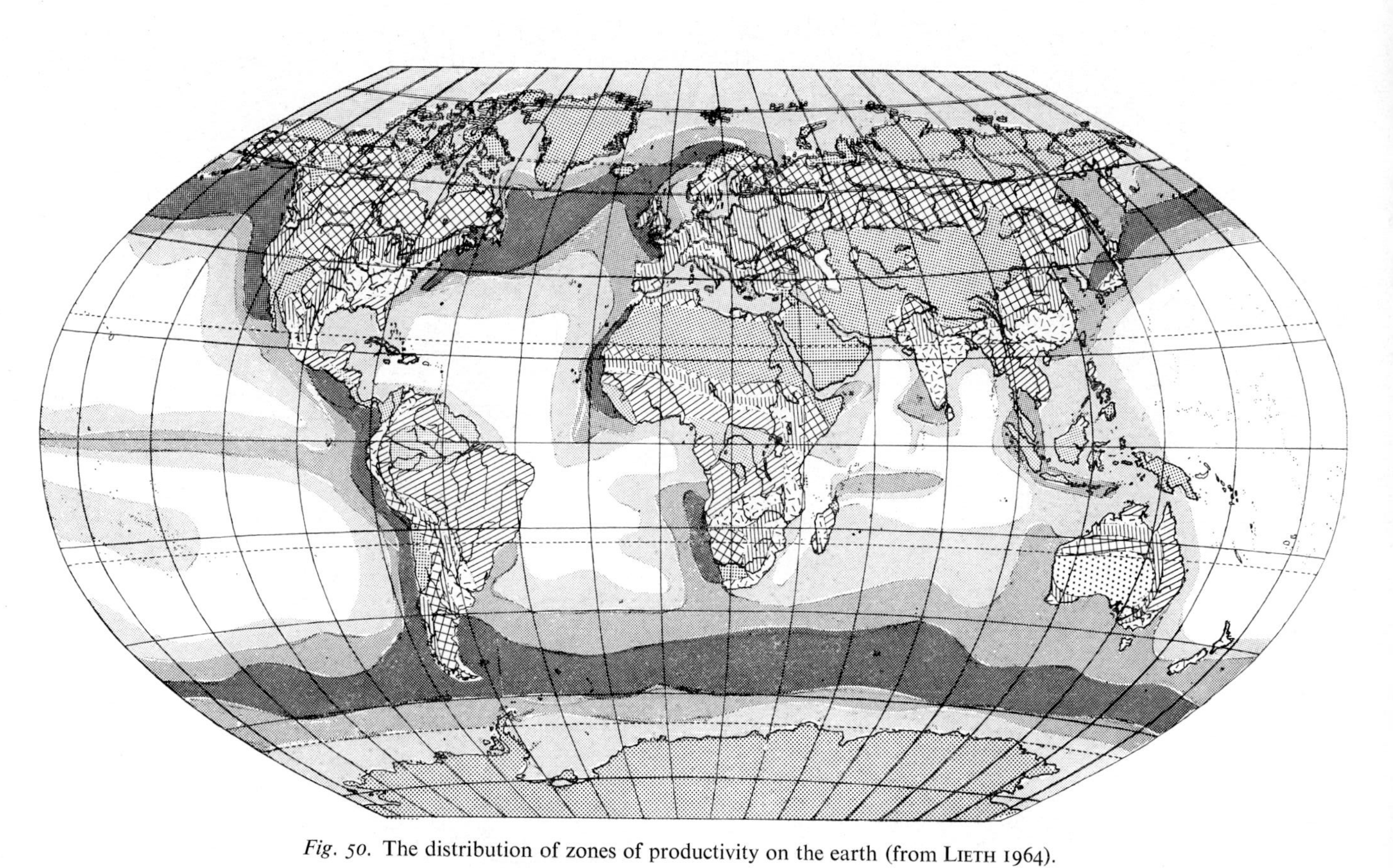

Fig. 50. The distribution of zones of productivity on the earth (from LIETH 1964).

acteristic adaptations to a subterranean life. The Australian pouched mole, (*Notoryctes typhlops*), the African golden mole (*Chrysochloris capensis*), the European mole (*Talpa europaea*) and the common skink (*Scincus scincus*) of the north African semi-desert all belong to different systematic groups. They nevertheless agree remarkably in the form of the head and limbs (SPATZ & STEPHEN 1961).

Certain desert snakes, including the North American rattlesnake *Crotalus cerastes*, the Sahara vipers *Cerastes vipera* and *Cerastes cerastes* and the Namib viper *Bitis peringueyi* move by sidewinding as an adaptative design for open landscape. Other extremely interesting adaptations shown by these snakes are burial in the sand, the displacement of the eyes on to the upper surface of the head and the disappearance of the snakes' characteristic hissing as a warning sign. Instead of the hiss, the scales are rubbed together.

The number of vegetational formations corresponds to the number of biomes (SCHMITHÜSEN 1968). To give an idea of their diversity some of the more extreme biomes will be discussed. The quantity and temporal distribution of precipitation is critical to the distribution and areal extent of biomes.

Vegetational and climatic zones of the tropics and the distribution of biomes.

Number of arid months	Division of lowland zones according to length of dry periods	Distribution of the commonest, widely distributed biomes in the climatic zones of the tropical lowlands	
1 2	Tropical rain-forest zone	Tropical rain-forest biome	Biomes of flood savanna with evergreen gallery forests
3 4	Wet savanna climatic zone	Monsoon forest biome	
5		Campo cerrado biome	
6 7 8	Arid savanna climatic zone	Thornbush and succulent biome	Arid savanna biome (without evergreen gallery forests) with dry deciduous forest or thorn trees.
9	Thornbush savanna climatic zone	Thornbush savanna biome	
10 11	Semi-desert climatic zone	Semi-desert biome	
12	Desert climatic zone	Desert biome	

The Desert Biomes

Desert and semi-deserts are biomes that produce some extraordinary forms of life (MALOIY 1962). Desert and semi-desert cannot always be sharply separated from each other, and indeed the semi-deserts usually form a frame surrounding the arid central areas. Edaphic and major climatic factors control the existence of desert conditions which are most widespread in the continental arid zone of the northern hemisphere of the Old World. The environmental conditions of desert biomes are characterised by: eleven or twelve arid months in the year; irregular precipitation usually totalling less than 150 mm *per annum*; sharp contrasts of temperature between day and night (+56°C to −40°C); high evaporation by the desert winds (such as the samum, harmattan, ghibbi and sirocco); and a predominance of mechanical weathering.

According to the predominant substrata the following types of desert can be distinguished: 1. Rocky desert = Hamada; 2. Stony desert = Serir or Seghir; 3. Salt-clay desert = Sebcha, Takyr; 4. Sand desert = with dunes; erg or barchan; 5. Salt desert.

Characteristic zonal soils of the tropical and subtropical arid zone of the earth are: 1) the fine desert soils, including loose sand, dunes and takyr soils; and 2) the semi-desert soils, including brown semi-desert soils, sierozem, and red-brown soils; intra-zonal soils include solonchok and solonetz. The general picture of deserts only changes near sources of ground water such as oases. Desert and semi-desert occur in the New World in western subtropical North America (Sonora Desert, Mohave Desert, Death Valley, Baja California), on the South American Pacific coast of Peru (South of 4°S) and Chile, and in north western Argentina.

The Chilean and Peruvian desert is one of the largest plant-free areas of the earth. At the weather station at Arica only 4 years out of 39, during which observations have been made, had more than 2 mm precipitation. In the Old World the most extensive desert areas are in the Sahara, on the Arabian peninsula, in Asia (Gobi, etc.), in South West Africa between 18° and 28°S and in central and western Australia. The Australian deserts have more the character of semi-deserts for they mostly receive more than 100 mm precipitation per year. Spinifex semi-desert alternates with acacia-scrub and thin cassowary-scrub.

STOCKER (1962) has drawn attention to an ecological asymmetry in the arid tropical zone. He pointed out that towards the equator the savanna biome with isolated trees stretches right up to the edge of the desert. Towards the temperate region, on the other hand, desert passes into treeless semi-desert, at most with a few small bushes, or else into the steppe biome.

The flora, the fauna and the people that live in the desert are adapted to these desert conditions. The number of species, nevertheless, by comparison with, say, tropical rain forest is extremely small. Thus in the South American hamada 250 plant species were found in 100 000 km^2 and in the southern Tunisian

Sahara 300 species in 150 000 km^2. The species diversity increases considerably in the Central Saharan Mountains, which ecologically are less uniform. Thus in Hoggar there are 350 species in 150 000 km^2; in Tibesti 568 species in 200 000 km^2; in Aïr 430 species in 150 000 km^2 and in Ennedi 410 species in 150 000 km^2. In the driest and most uniform areas of the Sahara the species-diversity for the same area falls below 50. Thus in Djourab there were 50 species in 150 000 km^2; in Ténéré 20 species in 200 000 km^2 and in Majabat 7 species in 150 000 km^2. Plants which can tolerate the extraordinary living conditions in the central Sahara include *Aristida pungens*, *Anabasis aretioides* (a hamada species) and *Genista saharae*.

The following groups of plants adapted to desert life can be recognised:

1) Rain plants. The desert springs to life and breaks into flower after rain – *Mesembryanthemum*, *Mollugo*, etc.

2) Poikilohydrous plants. These are plants whose leaves appear to be completely withered during drought, but on being wetted for a brief period, immediately turn green. Examples are *Selaginella* and *Cheilanthes*.

3) Xerophytic plants. For example *Aristida pungens*, adapted to sand desert.

4) Perennial plants, often with root systems reaching down to ground water. Examples are the tamarisks with roots extending 30 m deep.

5) Succulent plants with water-storing tissues. Examples are the cactuses, which are originally American and the Old World Euphorbiaceae.

6) Sclerophilous plants, with hard leaves or none, and with a covering of thorns.

7) Salt plants, including the Chenopodiaceae which occur in dry basins lacking any outflow.

Desert animals are adapted to desert conditions and to desert plants. Typical desert animals of the Sahara are, for example: the desert rats *Jaculus jaculus*, *Gerbillus campestris*, *G. nanus*, *G. gerbillus*, *G. pyramidum*; the desert fox or fenec; the mastigure lizard *Uromastix*, which is provided with salt glands; the vipers *Cerastes cerastes* and *Cerastes vipera* with their side-winding locomotion; the skink *Scincus scincus* (sand-fish); the desert lark, *Ammomanes deserti*, with desert coloration; the flightless desert locusts; beetles of the family Tenebrionidae, with their long legs, in 99% of cases flightless; and the 2 cm long desert wood-louse (*Hemilepistus reaumuri*).

Species belonging to the desert and semi-desert of North Africa in most cases extend into the Indian arid areas. Species or genera with this type of distribution are called Saharo-Sindhian. Examples of this are beetles of the genus *Mesostena*. It is said of these that: 'like the dromedary they are linked in distribution with the salt steppes and deserts of the west and central eremial zones. Primarily they belong to the salt steppes and perhaps partly also to the arid steppes. They only secondarily penetrate into the areas of pure desert.'[1]

1 'wie das Dromedar an die Salzsteppen und Wüsten der west- und mitteleremischen Zone gebunden sind, wobei sie primär den Salz-, teilweise vielleicht auch Trockensteppen eigen sind und erst sekundär in die reinen Wüstengebiete transgredieren'.

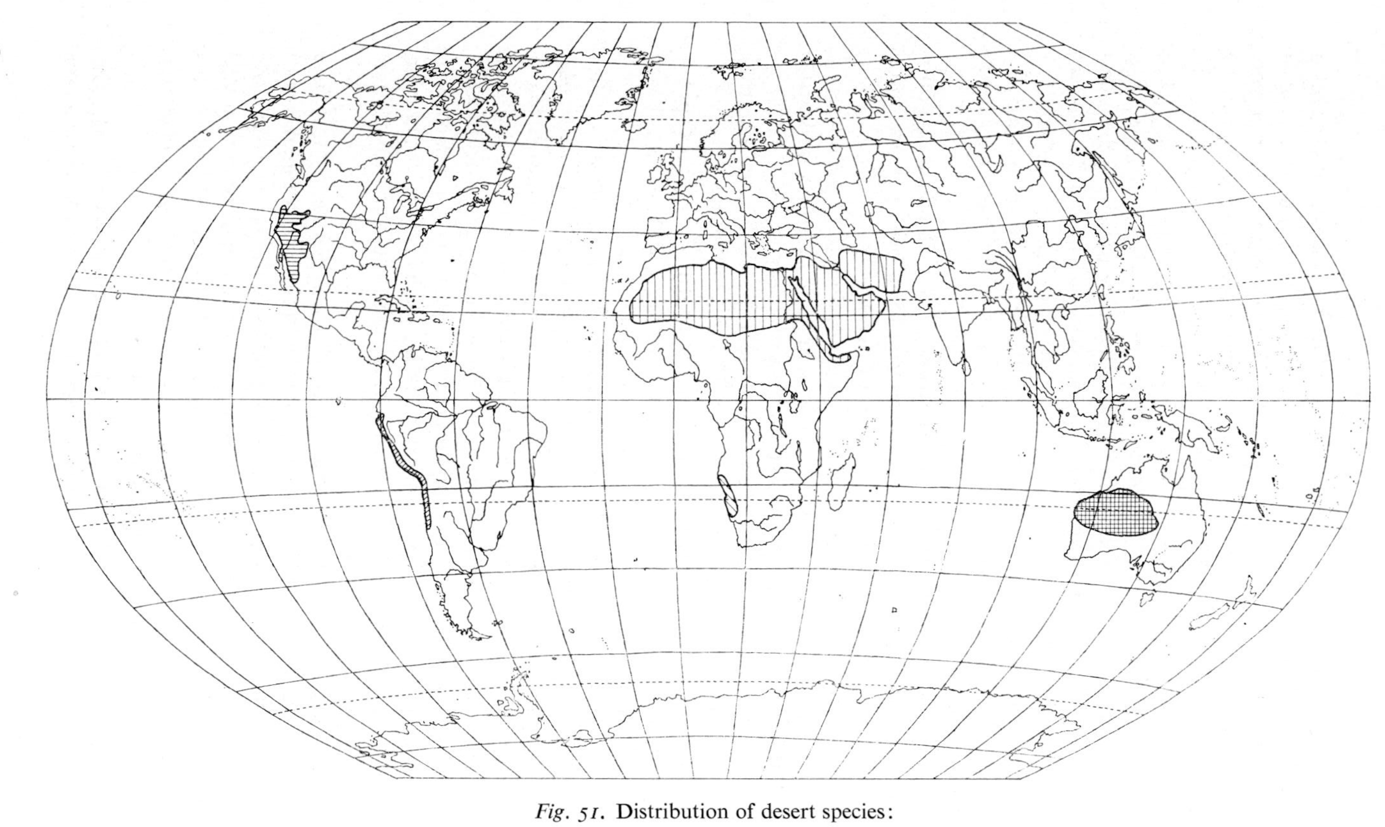

Fig. 51. Distribution of desert species:

horizontal lines = *Crotalus cerastes*; vertical lines = *Ammomanes deserti*; cross-hatching = *Stipiturus ruficeps*.

diagonal lines (/) = *Tropidurus peruvianus*; diagonal lines (\) = *Ammomanes grayi*;

The Saharo-Sindhian flora and fauna represent a biogeographical unity. In the semi-deserts of Rajastan mammal genera are characteristically met with that are otherwise known in the Sahara (e.g. *Gerbillus*).

The South West African Namib Desert is divided by the Swakop river into a southern section dominated by migrating dunes and a northern portion of stony desert. It contains endemic forms and also relatives of the Saharan fauna with corresponding adaptive types. WAIBEL (1921) said this about the Namib: 'The third day of creation, when God created 'grass, the herb yielding seed and the fruit tree yielding fruit after his kind' passed the Namib by. This South West African coastal desert seems like a vestige from the days of the earth's creation.'[2]

The miracle plant of the Namib – *Welwitschia mirabilis* – is a gymnosperm known from as early as the Miocene. It is named in honour of its discoverer, the Austrian doctor WELWITSCH. It is extremely well-adapted to life in the desert by its long tap root and its squat trunk, by having two leathery perennial leaves that only grow at the base, and by its dioecious reproductive parts arranged in cone-like masses: it is as well-adapted as the pebble or window plants of the genus *Lithops*, the sand-inhabiting reptiles such as *Palmatogecko rangei*, *Ptenopus garrulus*, *Aporosaura anchietae*, *Meroles cuneirostris*, *Meroles reticulatus*, *Acontias lineatus* and *Bitis peringueyi* (cf. LOUW 1972) or the reptiles occurring in the stony desert such as *Eremias undata*, *Rhoptropus afer*, *Rhoptropus bradfieldi*, *Cordylus namaquensis* and Tenebrionids (c.f. HOLM & EDNEY 1973). The burrowing gecko *Ptenopus garrulus* which inhabits dunes, is particularly worth noting, having adopted a subterranean mode of life.

The South American Atacama desert has much in common with the Namib. Both are coastal deserts resulting from cold sea currents. Landward winds lose their water by precipitation in passing over these cold currents. Then they warm up again over land and become under-saturated. In both of these coastal deserts terrible fogs occur without any appreciable rainfall; in South America they are called Garua. In many areas there are no plants at all.

In places in the Atacama desert there are 'islands' of *Tillandsia* species (Bromeliaceae) and of cactuses such as *Eulychnia*, *Copiapoa*, *Philicopiapoa*, *Mila*, *Borzicactus* and *Weberbauerocereus*. Among the animals the Agamidae of the Old World desert areas are completely lacking. They are replaced by the leguans (Iguanidae, including *Ctenoblepharis adspersus*) and the Teiidae including *Dicrodon guttulatum* and *D. heterolepis*. Viperidae are also absent from the New World. Instead there occur the fer-de-lance snakes of the genus *Bothrops*.

Although the cold Humboldt current is the cause of the desert conditions of the Atacama, in the sea it results in an extraordinary productivity of life. An enormous plant biomass of algae results in a great diversity of animal life,

2 'den dritten Schöpfungstag, da Gott Gräser und Kräuter schuf und fruchtbare Bäume, den hat die Namib nicht mehr miterlebt. So erscheint uns diese südwestafrikanische Küstenwüste als ein Relikt aus den Schöpfungstagen der Erde'.

dependent on various food chains on this plant biomass. Fish of the genus *Sardina* are specially important for the sea birds of Pacific South America. The faeces of the sea birds that live on these little fishes have piled up into a 60 m thick layer of guano on certain islands such as the Peruvian Chincha group. Guano occurs on the bird islands and bird mountains on the rainless subtropical west coasts of continents, especially in Peru, Chile and South West Africa.

The yearly increase in guano is estimated at 200 000 tons. Because of its high phosphate content it is in great demand as manure. ALEXANDER VON HUMBOLDT and J. LIEBIG were the first Europeans to draw attention to its value as fertiliser. The Arabs, however, manured their vines and palms with guano from the Bahrein Islands as early as the twelfth century, while the use of guano as manure was already known to the pre-Incan Chimus on the Peruvian coast (NIETHAMMER 1969). Ten million tons of guano was extracted on the Chincha Islands in the short period from 1851 to 1872. On many Pacific islands the coral limestone has been dissolved by bird dung and rainwater and converted into phosphate; an example is the coral island of Nauru (0°32'S; 166°15'E) which has an area of 32 km and rises from a sea floor about 2 000 m deep. The phosphate production of Nauru, which is dependent on an animal food chain, amounts to two million tons per year. It has allowed the inhabitants to pass social legislation of a type achieved in few lands on earth; they have a uniform monthly rent of about £1.30, free education, telephones and hospitals etc.

Two types of guano can be distinguished on the basis of colour, age and content of phosphoric acid. These are:

1) Red, or fossil, guano with 20–30% of phosphoric acid

2) White, or recent, guano with 10–12% of phosphoric acid; 10–12% nitrogen and 3% potassium.

The most important guano birds of South America are: the guano-cormorant *Phalacrocorax bougainvillei*, which is the breeding bird of the Buenaventura Bay of Colombia, and of the Peruvian Coastal Islands south to the Mocha islands of Chile; *Pelicanus occidentalis*; and several species of boobies (*Sula nebouxii*, *S. variegata*, *S. dactylatra*). On the southern island of the Chincha group alone the colony of *Phalacrocorax bougainvillei* contains about 360 000 individuals and occupies an area of 60 000 m^2, i.e. about 3 nests per m^2. The penguin *Spheniscus humboldti* makes its nesting burrows in the guano.

In general it is only birds of the order Pelecaniformes which are guano birds, but there are some notable exceptions. One such is the sooty tern (*Sterna fuscata*) on the Atlantic island of Ascension. Artificial islands off the South West African mainland, in the form of platforms made of planks, accumulate 6 cm of guano yearly (NIETHAMMER 1969).

Less economically important is the guano dropped by bats in caves, and seal guano.

The Australian deserts or Eremaea lie on the same latitudes as the Kalahari. They are limited to the south by the Nullarbor plain which stretches 700 km from east to west and is formed of Tertiary karst limestone. In the sand and clay plains there is a saltbush semi-desert with the two dominant species of *Atriplex vesicaria* (saltbush) and *Kochia sedifolia* (bluebush). Between these are islands of the hard leafed spinifex grass (*Triodia pungens*, *Plectrachne schinzii*) and the mulga scrub with *Acacia aneura*, *Cassia* spp and small species of *Eucalyptus*. The characteristic fauna of the Australian deserts includes jerboa marsupial mice (*Antechinomys laniger*, *A. spenceri*), the western hare wallaby (*Lagorchestes hirsutus*), various birds (*Polytelis alexandrae*, *Stipiturus ruficeps*) and numerous reptiles. The Agamid lizard *Amphibolurus isolepis* lives in the deserts of west and central Australia and feeds mainly on ants, other arthropods and plants. The species is ecologically strictly limited to sandy soils and is absent in the interspersed rocky plateau areas. PIANKA (1972) studied the Australian desert lizards and was able to show that: 'Sandridge, shrub-*Acacia* and sandplain-*Triodia* habitats are three particularly important habitats to which lizards have become specialized.' In 72 species out of the total of 94 that occur in the Australian desert areas he was able to establish the following adaptational or range types:

7 ubiquitous species
8 northern species
10 southern species
6 Eremaea species
7 central relicts
8 sandridge species
16 shrub-*Acacia* species
10 sandplain-*Triodia* species

---

72 Total

By contrast with the desert biome, the tropical rain forest or hylaea is a biome with extremely high plant productivity. The major climatic conditions for this biome can be defined as: a uniform day and night temperature of 25–30°C with complete absence of frost; and uniformly high precipitation, usually above 2000 mm per annum. The altitude limit for the biota of the tropical rain forest lies in general at a height of 1500 m in the mountains. At this height the annual temperature is only 15 to 20°C, while at 3000 m frosts occur. As an adaptation to these changes with altitude the flora is divided into different altitudinal zones (stratification).

Because of the lack of any marked annual variation in climate there are usually no annual rings in the cross-sections of tree trunks. The giant trees of

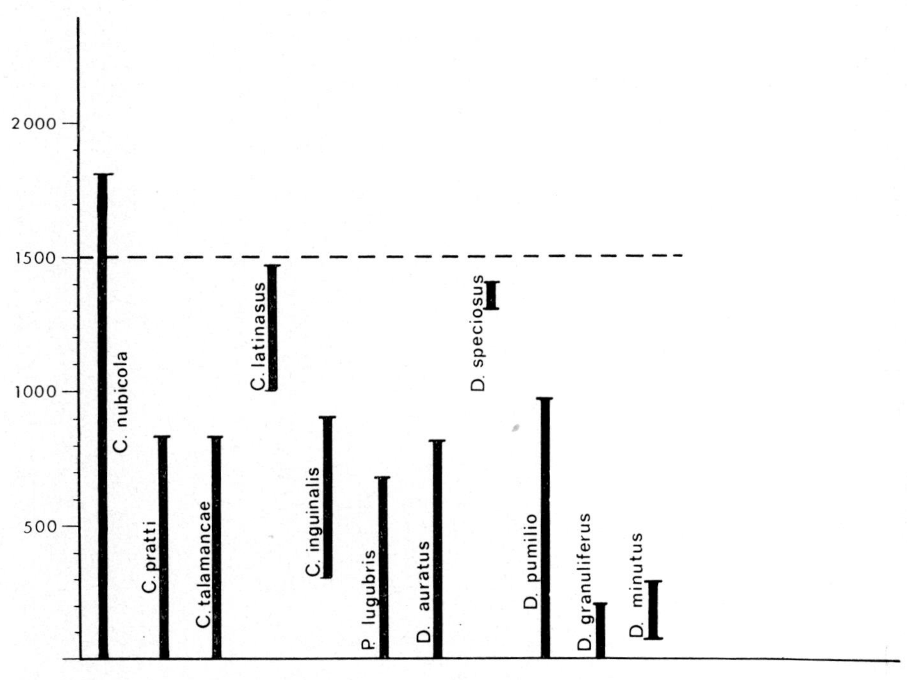

Fig. 52. Vertical distribution of Central American Dendrobetidae (after SAVAGE 1968).

Fig. 53. The rain-forest biom (Island of Florianopolis, Santa Catarina, Brazil, march 1969)

the rain forest are held up by buttress roots and their leaves mostly have entire margins and rain points. The period of endogenous leaf change varies between seven and twenty months. Leaf development is very quick and could be described as a sudden unfolding. Flowers often grow directly on the main stems of the plants (cauliflory) and the seeds, by contrast with desert plants, do not long remain fertile. The tropical lowland rain forest is rich in fruit trees. 90% of all liana plants inhabit it. Just above soil level there is a high concentration of CO_2 (0.34% in south east Brazil according to MACLEAN; in general 0.2% according to BÜNNING) and also high humidity (more than 90%) and low illumination. The plants have special shade adaptations. The rain forest is usually developed on laterites, which are red, humus-poor soils with only a small quantity of food material for plants. The soil cover itself is usually formed from yellow or reddish yellow latosols. Rain forest soils that have been exposed by shifting cultivation have usually lost all their organic material by leaching, and bracken (*Pteridium aquilinum*) usually spreads on the devastated surfaces.

A striking feature of tropical rain forest is the proportion of phylogenetically

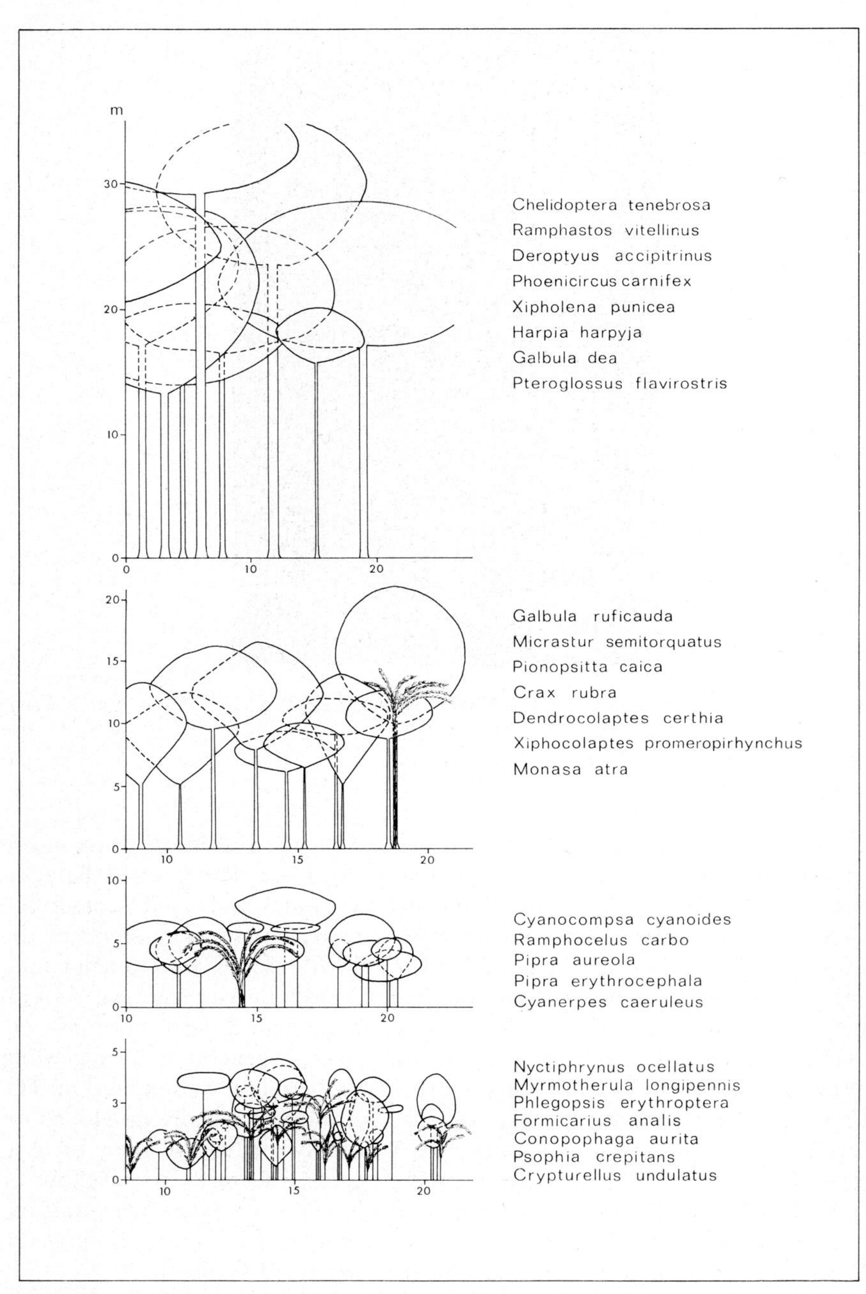

Fig. 54. The vertical structure of the tropical rain forest (after MÜLLER 1973).

The Savanna Biome

The savanna concept is derived from the Spanish word sabana, meaning grassland. Nowadays it is used exclusively for the grasslands of the seasonally rainy tropics regardless of how they arose or to what extent they are interspersed with trees or bushes. There are three different types of savanna which are controlled by major climatic conditions and are characterised by particular types of soil. These types of savanna are:

a) Humid savanna and campos cerrados, with 3 to 4½ arid months per year.
b) Arid savanna, with 6–7 arid months per year.
c) Thornbush savanna with 8 to 9½ arid months per year.

Within the tropics the savanna is an abrupt contrast to the rain forest. 'The contrast between these two different worlds is found through the whole of tropical Africa. The dark hostile forest comes up against the open clear grassland ... On the high savannas nature seems more under control. The captivating beauty of the forests is lacking, but so too are the hostile powers and demonic forces. A man can breathe under the open sky, the breast swells comfortably and courage and strength return. There are dangers here as well, but they come on us openly, without deception or malice. One can guard against them and is not so helplessly abandoned to them as down in the witch's cauldron of the forests. This important change in the character of the landscape is controlled by the occurrence of grass. The open outlook, clear sunshine and brighter mood all go back to grass.'[1] (WAIBEL 1921)

In the humid savanna precipitation is still more than 1,200 mm per year. During the rainy season the ground is covered 100% with plants. The proportion of 'fire-proof' trees and pyrophytes is large. These reach a height of 6 to 12 m on average and they are rain-green, large-leaved and thick-barked. The gallery forests that occur in humid savanna remain untouched by forest fires.

1 Durch das ganze tropische Afrika geht der Gegensatz dieser beiden verschiedenen Welten. An den dunklen feindlichen Wald grenzt das offene, heitere Grasland... Auf den hochgelegenen Savannen erscheint uns die Natur heimischer; es fehlt die sinnberückende Pracht der Wälder, es fehlen aber auch die feindlichen Mächte, die demonischen Gewalten. Man atmet auf unter dem freien Himmel, es dehnt sich wohlig die Brust, Mut und Kraft kehren zurück. Auch hier oben drohen Gefahren; aber sie treten uns offen entgegen, ohne Hinterhalt und Tücke. Man kann sich ihrer erwehren, ist ihnen nicht so hilflos preisgegeben wie da unten im Hexenkessel der Wälder. Dieser bedeutende Wechsel des Landschaftscharakters ist durch das Auftreten des Graswuchses bedingt. Der weite Blick, der heitere Sonnenschein, die gehobene Stimmung, all das geht auf das Gras zurück'.

Fig. 61. Distribution of typical savanna species;

||||||||| = *Crotalus durissus*, ≡ = *Psephotus varius*, ////. = *Nectarinia senegalensis*.

In the arid savanna precipitation is only between 500 and 1,100 mm per year. Older authors referred to many types of arid savanna as orchard steppe, because of their interrupted tree cover. The grass cover is 1 to 2 m in height and consists of separate, hard-leaved tufts which no longer form a continuous turf. Trees are sometimes lacking completely in the arid savanna, but when they do exist they are 5 to 10 m high and belong to the growth types of the arid deciduous forest. Lianes are absent from the arid savanna.

In the thorn-bush savanna (thornbush steppe) precipitation is only 200 to 700 mm per year. The grass cover is only 30 to 60 cm high, in correspondence with the large number of dry months in the year (8 to 10). Thornbushes predominate, such as Acacias and Mimosas. Trees with bark assimilation and succulent stems occur, such as the baobab tree (*Adansonia digitata*) and the bottle tree (*Adenia globosa*). In the Australian brigalow scrub there are Acacias (Brigalow = *Acacia harpophylla*), the custard apple (*Brachychiton rupestris*) and species of *Eucalyptus*. In the Brazilian caatinga there is a thorn-savanna forest with Bromeliaceae and cactuses.

A special fauna is adapted to this savanna formation. In the African and Indian savannas hoofed mammals are characteristic of this type of landscape. The East African savanna has zebras, giraffes, hartebeest, topi, waterbuck, impala, giraffe-necked gazelle, Thompson's gazelle, Grant's gazelle, African

Fig. 62. Zebras in the African savanna (SERENGETI, 1972).

buffalo, the black or hook-lipped rhinocerous, elephants, spotted hyaena, jackals, Cape hunting-dogs, cheetahs, ostriches, secretary birds and puff adders. By contrast with the African savanna, the Indian savanna is much poorer in species, although most of the African savanna fauna is of Oriental origin, as shown by fossils, especially those of Sivalik.

In the Australian savanna there is a predominance of marsupials, parrots, parakeets and waxbills. The African ostrich is replaced in Australia by the emu.

The South American savanna (cf. campo cerrado), like the Australian, is very poor in large animals. Nandu, maned wolf, pampas deer and the cariama should be mentioned. The Old World vultures of Africa and India are here replaced by the New World vultures.

Two particular types of savanna are flood savanna and termite savanna. Flood savanna has a grass cover up to 3 m high which is inundated once or twice a year. Natural raised river banks usually carry trees – raised bank forests or banco-forests. These are evergreen or partly deciduous stands, often with palms. In South America these are especially *Mauritia vinifera* and *Copernicia*; in Africa they are species of *Hyphaena* and deleb palms (*Borassus*).

In flood savanna the climate determines the environment of the animals, while in termite savannas the animals determine the nature of the landscape, much as

Fig. 63. Migrating gnus in the Serengeti savanna (SERENGETI, 1972).

in coral reefs. Termites, and also the leaf-cutting ants of South America, allow the regular occurrence of damp-loving woods because of their burrows and the way that they work over the soil. HESSE's view (1955) that these islands of forests are only formed on abandoned termite hills can no longer be maintained as a generalisation. The great majority of all known termite species – about 1900 in all – live in the tropics and sub-tropics (EMERSON 1952). 41 species are however known in the Palaearctic region. Africa has the greatest diversity with 570 species and 89 genera. In passing it is noteworthy that many of these species live in true rain forest, unlike the Australian situation. The savanna and rain forest species build their nests different distances apart from each other, according to species. LEE & WOOD (1971) give an interesting table concerning this:

Abundance of termite hills

Species	No. of hills hectare	Occurrence
Anacanthotermes ahngerianus	162	Central Asiatic steppes
Coptotermes lacteus	1–2	Arid forest of southern Australia
Odontotermes sp.	5–7	Kenya savanna
Macrotermes bellicosus	2–3	Congo savanna
Macrotermes spp.	3–4	East African savanna
Nasutitermes exitiosus	4–9	Arid forest of southern Australia
N. triodiae	3–7	Tree savanna of northern Australia
N. magnus	61	East Australian pasture land
Trinervitermes trinervoides	435	Southern African savanna
Amitermes laurensis	28–210	Savanna forest of northernAustralia
Drepanotermes spp.	more than 350	Semi-arid forest of Australia
Cubitermes fungifaber	875	Tropical Congo rain forest
C. exiguus	0–652	Congo savanna
C. sankurensis	8–550	Congo savanna

However it is important to emphasize that not every species can produce termite savanna (TROLL 1936). FULLER (1915) established that *Macrotermes natalensis* in particular was the species characteristic of parklands of this sort in Natal. In Africa species absent from the rain forest include *Macrotermes falciger*, *M. bellicosus*, *M. subhyalinus* and *M. natalensis*.

The savannas are the areas where swarms of the migratory locusts are found. To control these migratory locusts Argentina paid, on average, between 1897 and 1935 £300,000 sterling per year, the Union of South Africa paid £140,000 sterling in 1934 alone, and the U.S.A. from 1925 to 1934 paid out $ 4,500,000. In September 1948 more than 10,000 tons of rice was destroyed by locusts. As a result of the founding of the Anti-Locust Research Centre in London and in close collaboration with the FAO, world-wide agreements on the control of locusts were reached in 1946 and 1948.

UVAROV (1951) has been able to show that every migratory locust has a migratory phase (*phasis gregaria*) and a solitary phase (*phasis solitaria*). These differ from each other in morphology, physiology and behaviour and can be converted into each other experimentally. The occurrence of the migratory phase in *Locusta migratoria* and *Schistocerca gregaria* has in the past resulted in great losses. The migratory phase is not caused by lack of food, but by an increased rate of reproduction and consequently a higher population density. These in turn are connected with higher rainfall near the limits of geographical range, are favoured by high temperatures when the atmospheric pressures decrease and are controlled by the glands known as corpora alata. Migration begins with the larvae of the *gregaria* phase, which move forward at a rate of 300 to 350 m per day. Migration continues in the adult to a greatly increased extent. The moment of a locust outbreak depends on sun, temperature and wind. Huge swarms are formed which drift, largely passively, before the wind. The flight performance of the individual locusts during these flights is very well adapted to the wind speed.

While *Locusta* covers huge distances, *Calliptamus* and *Dociostaurus* only undertake smaller migrations. The descent of migratory locusts usually happens during rain. The animals begin to eat, sometimes lay eggs to the number of 70 to 100 per female, and then migrate again. The particular populations of *L. migratoria* which arise in this manner can maintain themselves for a long time

Fig. 64. Flood savanna with water buffalo. (The island of Marajó, Amazon basin, 1969).

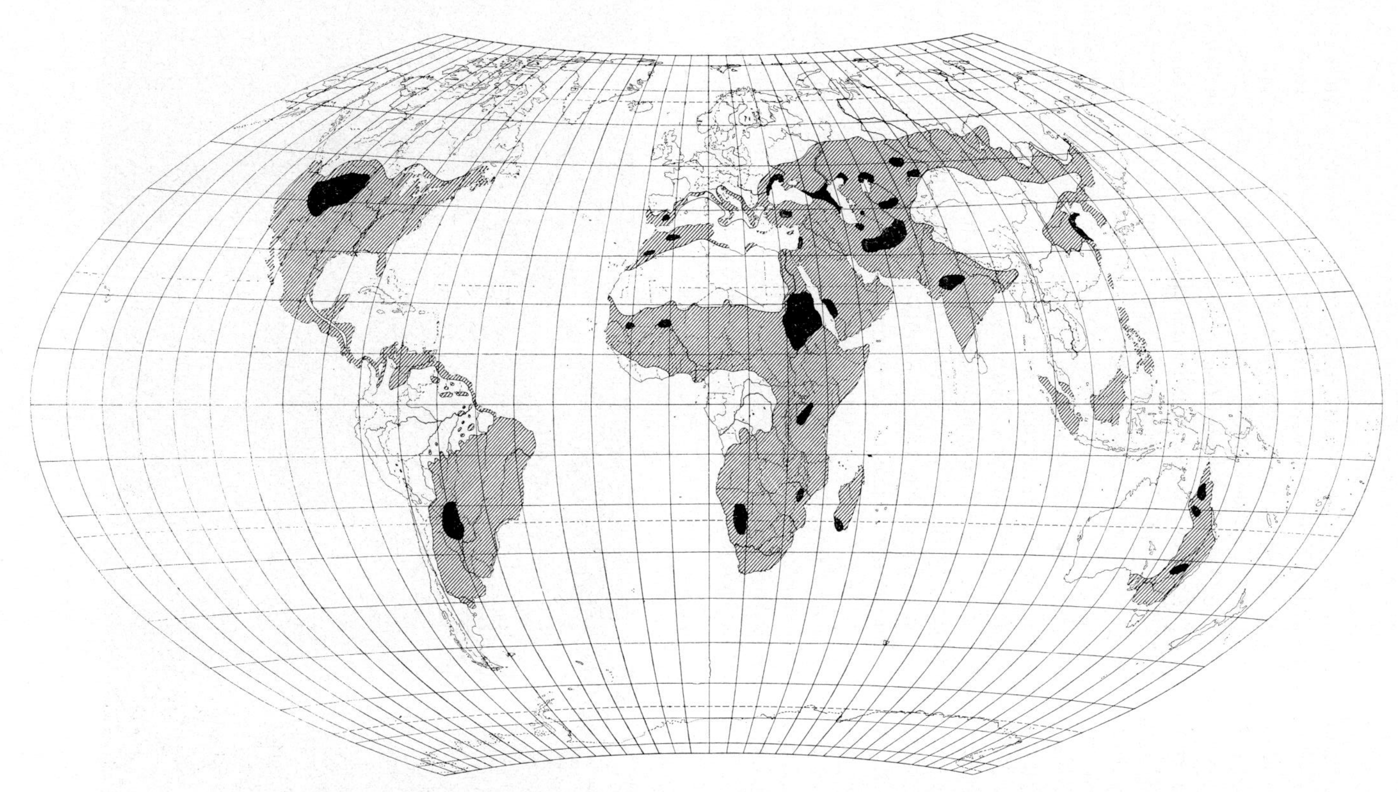

Fig. 65. Areas of distribution (shaded) and of swarm formation (black) of the most important species of migratory locust (from MÜLLER 1972).

if the circumstances are right. Given a change in wind direction, even a return migration is possible. Thus a big migration of *Melanoplus mexicanus* took place in 1876 from the Rocky Mountains in the direction of the Mississippi with a mass return flight in 1877, as described by PACKARD & THOMAS (1878) and PACKARD (1880). In the last few decades calamities have occurred in southern Europe as a result of the locusts *Calliptamus italicus* and *Dociostaurus maroccanus*. (In June 1930 a swarm of *Calliptamus* near Vienna actually stopped a railway train). Europe is not in danger, however, from *Locusta migratoria*. The outbreak areas of *Locusta* lie in uncultivated areas, while those of *Calliptamus* are in cultivated ones. Even as late as 1947, 300 hectares of cultivated land were damaged by *Calliptamus* and in June 1930 there was a mass reproduction of the species on the Griesheimer Sand near Darmstadt.

The distribution of many causative organisms of human diseases is controlled by the belt-like arrangement of the savanna biome, just as changes in the savanna biome control the dynamics of the locust populations. Louse relapsing fever spread in the African savanna areas at the beginning of this century. Knowing the ecological valency, the directions of migration and infestation of the disease could be predicted. The carriers of the South American Chagas' disease are assassin bugs (*Triatoma infestans* and *Panstrongylus megistus*) which are ecologically strictly confined to the campo cerrado biome. This campo cerrado is a special case which has been placed by manyscientists in the "rain-green" forest,

Fig. 66. The Campo Cerrado (near Pousada do Rio Quente, Goias, Brazil, march 1969).

regarded by others as a type of savanna and finally assigned by SCHMITHÜSEN to the open forest formation.

The campos, from the Portuguese word for field or open land, are well-lit, poorly shaded biomes of tropical South America. According to their varying tree cover, which is generally from 3 to 8 m high, they can be divided into campos cerrados (Port. *cerrado* – dense or closed), campos sujos (Port. *sujo* – dirty) and campos limpos (Port. *limpo* – pure or free). Campos sujos are campos in which stretches of open grassland predominate because of great distance from groundwater or human influence; individual trees and stands of bushes are nevertheless present, unlike campo limpo. The greater part of the Central Brazilian campos lie in the high-summer-rainfall areas of the tropics, with 7 to 8 wet months and more than 1300 mm precipitation. In the immediate proximity

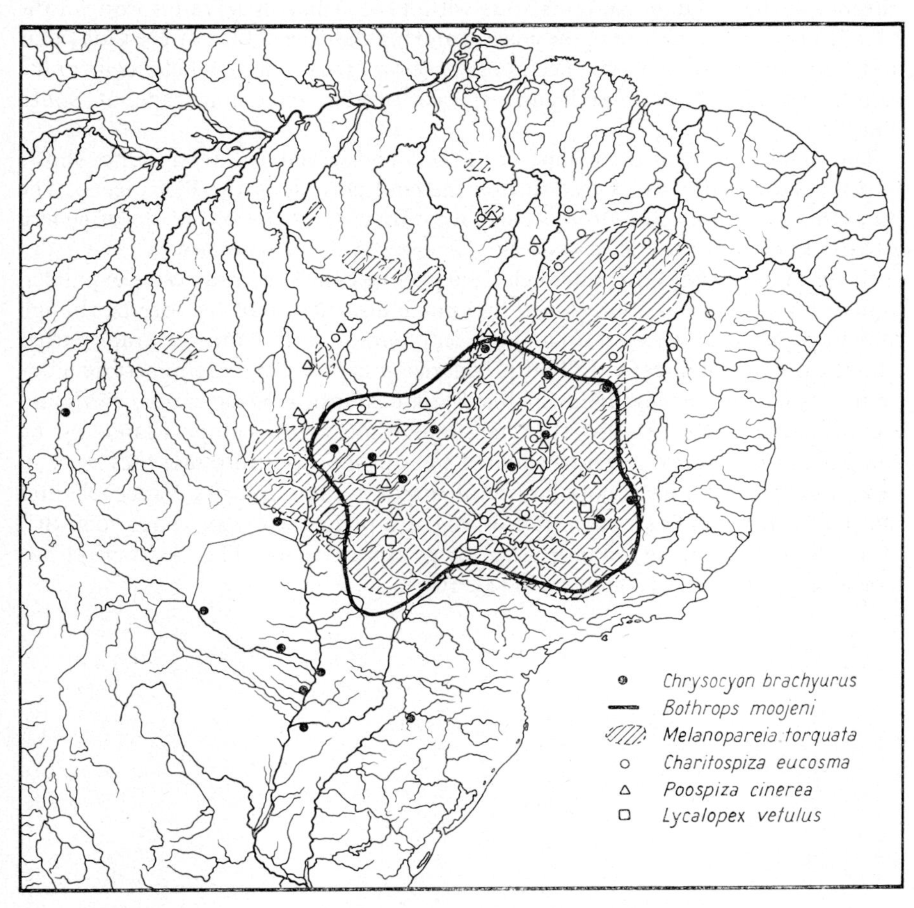

Fig. 67. Ranges of faunal elements of the Campo Cerrado (after MÜLLER 1973).

of ground water stands of trees 12 to 15 m high are produced and in water-bearing valleys there are canyon forests, mostly evergreen. The high proportion of grasses has caused many authors to equate the campo cerrado with the African savanna, but Waibel, who knew the African savannas by personal observation, already rejected this. In the campo cerrado there are no herds of large animals. Characteristic animals of the campo cerrado are the birds: *Melanopareia torquata*, *Cariama cristata* and *Rhea americana*; the wild dogs *Lycalopex vetulus* and *Chrysocyon brachyurus*; the poisonous snakes *Crotalus durissus* and *Bothrops moojeni*; the ground leguan *Hoplocercus spinosus*; and the frogs *Bufo paracnemis* and *Hypopachus muelleri*. The closest relatives of the campo cerrado fauna occur in the north west Brazilian caatinga region, in the chaco, on the isolated high campos, in the *Araucaria* forests of the state of Parana, in the isolated campo islands within the Amazonian rain forest, in the Llanos of the Orinoco and in the coastal and highland savannas of the Guayanas and Venezuela. Of the eight species of tinamou (Tinamidae) which occur in the campo cerrado, five are indicator species of the open landscape of South America.

The fossil history of the campo cerrado, as preserved at Lagoa Santa, shows that this biome already existed during the cold periods of the Pleistocene. The same thing is indicated by the occurrence only in the campo cerrado of genera already known in the Pleistocene eg. *Chrysocyon* and *Lycalopex*.

In very recent times the faunal elements of the campo cerrado have expanded in distribution. This is conditioned by human destruction of the forests, though the campo produced by man has little in common with campo cerrado either climatically or botanically. In South America it is possible to demonstrate a dry period from 5000 to 2300 BC. which led to an extension of the campo areas and a contraction of the rain forests. The populations of the campo islands of the Amazonian rain forest are now isolated but during this arid phase they were in genetic exchange with the populations of origin in the Llanos of Venezuela and the central Brazilian campo cerrado. This arid phase gave way about 2000 BC. to a wet period during which the forest advanced again. This wet period has continued to the present day (MÜLLER 1973).

The Steppe Biome

In the steppe biome animals occur that are adapted to the extra-tropical grasslands, the prairies and the grasslands of tropical mountains such as the puna of the Andes. By contrast with the savanna the steppes experience a sharp change between cold and warm seasons. Characteristic soil types for the steppes are chernozems and chestnut soils. The East Patagonian steppe belt of South America, in the rain shadow of the Andes, is an arid steppe with tufty hard-leaved grasses and dwarf scrub. To the north it passes into semi-desert and to the south into deciduous forests of the southern beech (*Nothofagus*). Similar relationships occur on the lee side of the South Island of New Zealand. In the literature, especially on Africa, the expression 'steppe' has often been used in connection with savanna. This has led to a confusion of the two concepts. In the southern hemisphere there are certainly a number of marginal cases where steppe comes in contact with savanna. Examples are the grassland of the Ngorongoro crater and the equatorwards salient of the steppes into Argentina and Paraguay. In the northern hemisphere the relationships are considerably simpler.

Characteristic animals of the Eurasian steppe are ground squirrels (*Citellus*), the pallid harrier, the saker falcon, the steppe or tawny eagle, PANDER's ground jay (*Podoces panderi*), the lark *Melanocorypha mongolica*, the chigetai wild ass, and the saiga antelope. On the European steppe of the late Pleistocene lived the giant deer (*Megaloceros giganteus*). Characteristic species of the North American prairies are the prairie chicken (*Tympanuchus cupido*), various rodents (*Perognathus flavescens*, *Spermophilus franklinii*, *S. tridecemlineatus*, *S. spilosoma*), prongbuck (*Antilocapra americana*), bison (*Bison bison*) and rattlesnakes (*Crotalus*).

On the grasslands of the High Andes live the vicuña (*Lama vicugna*) and the guanoco (*Lama guanocoe*). The guanaco has also settled in the Argentinian pampas and on the cold steppe of Patagonia, where it occurs together with the armadillos *Zaedypus pichiy* and *Chaetophractus villosus*, the ground leguans *Proctotretus pectinatus*, *Liolaemus magellanicus* and *L. fuscus*, the wild dogs *Lyncodon patagonicus* and *Dusicyon griseus* and Darwin's rhea *Pterocnemia pennata*. The seasonal change between summer and winter affects the life of the steppe animals much more than those of the savanna. Steppe and savanna animals, being species of open landscapes, have excellent powers of sight and smell. Cursorial animals mostly form herds.

One particularly controversial steppe has been the South American pampa,

Fig. 68. The Puna-Biom in 4600 m (near Campamento Mina California/Argentinia, WERNER 1967).

Fig. 69. Esparto grass steppe in the Moulua basin (Maroc, 1972).

the agricultural heartland of Argentina. Some scientists held that this was originally forest land that had been laid waste by the bush fires of the Indians. CHARLES DARWIN (1832) was the first to consider the problem of the treelessness of the pampa. During his ride from Bahia Blanca to Buenos Aires, He distinguished very clearly between the original landscape of the pampa, which on evolutionary grounds he held to be originally treeless, and the plant growth which was now possible on the pampa. Later writers have often completely neglected to make this distinction. More recent biogeographical studies on Neotropical dispersal centres have shown that the pampa represents an inde-

Fig. 70. Types of distribution of species of the South American high steppe (Puna) and cold steppe.

pendent dispersal centre of the steppe fauna. This confirms the supposition that the pampa was originally grassland.

There are disease phenomena associated with steppe just as there are causal organisms and disease vectors associated with tropical rain forest or with savanna. JUSATZ (1966) has described an excellent example. The causal organism of Tularaemia is *Pasteurella tularensis*. It is transmitted by ticks and mites on rodents in the steppes of the USSR, on hares in the rest of Europe and on sheep in the USA. The occurrence of Tularaemia is associated with continental climates and with steppe land or pasture derived from steppe.

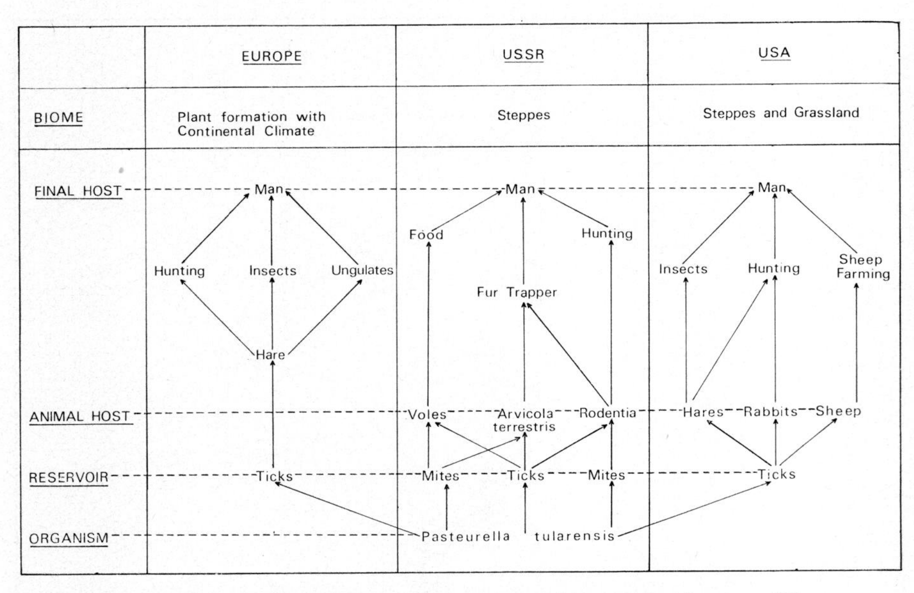

Fig. 71. Connection between *Tularaemia* and the steppe biome (after Jusatz 1966).

The Tundra Biome

The name tundra derives from the Finnish word *tunturi*, meaning an unforested hill.

The vegetation of the tundra biome is dominantly treeless. Mosses, lichens and heaths rich in low scrub predominate. In the northern hemisphere tundra biomes exist in the subpolar region of northern Eurasia, on the coasts of Greenland, in Iceland and on the margin of the North American Arctic (cf. WIELGOLASKI & ROSSWALL 1972). In the southern hemisphere they extend to much lower latitudes. The high oceanic climate and the small size of the Antarctic islands makes it impossible to decide for certain whether the southern tree and forest line is decided by lack of warmth or by the action of wind. The forest line, and therefore the beginning of the tundra biome, runs from the southern tip of South America (56°S), north of the Falkland Islands, reaching 40°S in the

Fig. 72. The tundra biome in Spitzbergen (Remmert 1973).

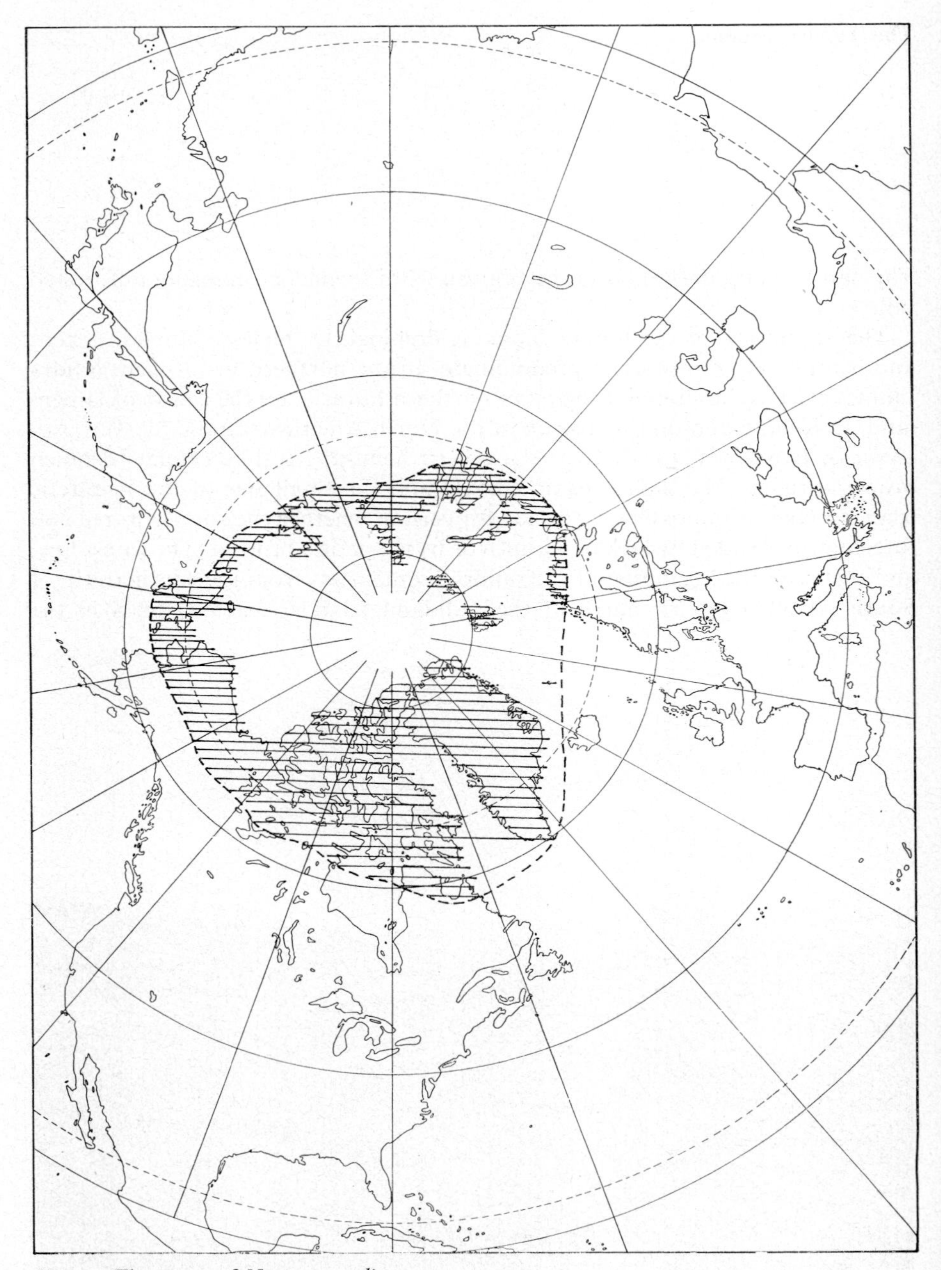

Fig. 73. The range of *Nyctea scandiaca.*

southern Atlantic Ocean. In the Indian Ocean, between St. Paul and New Amsterdam, it even reaches to 38.5°S. Tristan da Cunha, at 37°S in the Atlantic Ocean, has one species of tree.

Large and sharp increases of temperature such as occur in the high mountains, are not characteristic of the tundra biome in the summer months. By contrast with the fauna of high mountains, Arctic insects will not survive strong fluctuations of temperature. Their period of development is considerably lengthened because of the uniformly low temperatures. The very slow rate of plant growth is also a result of uniformly low temperatures. Thus a juniper trunk only 83 mm thick had 544 annual rings, and a spruce 60 cm thick had more than 400. The widespread reindeer moss, which in fact is the lichen *Cladonia rangifera*, grows 1 to 5 mm in a year. Tundra surfaces that have been grazed by reindeer therefore need at least ten years to recover.

The summers are brief and cool, for the mean annual isotherm is 0°C. The soil remains continually frozen at depth. The great annual oscillation in the distribution of radiation, from polar day to polar night, is critically important.

Characteristic animals of the tundra are: the reindeer (*Rangifer tarandus*); the blue hare (*Lepus timidus*); the arctic fox (*Alopex lagopus*); the musk ox (*Ovibos moschatus*) which now occurs only in Greenland and in Spitzbergen, where it has been introduced; lemmings, which account for the occurrence of northern birds of prey such as the rough-legged buzzard (*Buteo lagopus*) and the snowy owl (*Nyctea scandiaca*); the white grouse; and the snow-bunting.

The snowy owl is resident throughout the year in the Arctic, but the reindeer appears only in summer. In winter it moves southwards in gigantic herds.

The abundance of gnats and mosquitos characterizes the summer months in the tundra. It can be explained by the large number of pools of melt-water suited for reproduction and also by the fact that female midges can live on plant sap, without sucking blood.

In REMMERT's opinion (1972) the warm-blooded vertebrates are a basic cause of the preservation of the tundra. 'Their extinction would amount to the extinction of the tundra. And in that case the fate of the tundra will have a critical effect in winter on the European coasts to which the wild geese of the Arctic migrate.'[1]

Polewards of the tundra biomes stretch the arctic biomes. It is characteristic of them that their land fauna is critically dependent on the productivity of the sea, and is therefore essentially a coastal fauna. A good example of this is the polar bear which lives in the region of the north pole and feeds mainly onse als and water birds.

1 'Ihre Vernichtung würde einer Vernichtung der Tundra gleichkommen. Damit wird das Geschick der Tundra im Winter auch an den europäischen Küsten entschieden, wohin die Wildgänse der Arktis ihre Wanderungen unternehmen'.

The oreal biomes lying above the forest-line are closely related to the tundras. The altitude of the oreal centres can also be controlled by temperature factors. The ecological causes controlling the forest-line can be extraordinarily complex. Actual and potential forest-lines, in consequence, do not necessarily have to agree (cf. HAFFNER 1971, HOLTMEIER 1971, 1973, TROLL 1962, MEYER 1974).

Comparison of actual and potential forest-lines in the Alps

Station	Altitude (m)	Actual forest-line	Potential forest-line
Arosa	1865	1980	2250
Schatzalp	1872	2000	2287
Julier-Hospiz	2237	2050	2237
St- Moritz	1853	2120	2299
Bernina-Hospiz	2258	2180	2381
La Rosa	1873	2070	2243

By contrast with tundra the oreal biomes are much more independent of each other, in agreement with their isolated situations. Relationships with the tundra by no means exist everywhere. This is particularly true for the Mediterranean mountain ranges (VARGA 1970), the Andes (VUILLEUMIER 1970, MÜLLER 1973), the mountains of Africa (MOREAU 1966) and the Himalayas (DIERL 1970).

If the Arctic tundra is compared with the high meadow land of the Alps certain striking differences appear. Indeed the unification of oreal and tundral into oreotundral can only be justified to a certain extent, from the viewpoint of historical biogeography, in so far as it is justified at all. There are obvious differences between the two major environments in specific constitution and specific diversity. Thus REMMERT (1972) studied the tundra of Spitzbergen. He was able to show that certain types such as diplopods, wood-lice and snails, which in tropical and temperate latitudes were among the chief primary agents of decomposition, only made up 1% of the biomass in the tundra. This contrasted with the Alps, for example, where they made up 40% of the biomass.

The agreements between the tundras and the oreal biomes of Eurasia are of historical origin. At present there are numerous species that have a disjunct range with one population as a faunal element of the tundral and the other as a faunal element of the oreal. During the glacial periods the species of the European oreal and tundral occupied central Europe, producing a mixed fauna at that time (THIENEMANN 1914). Remains of this mixed glacial fauna were preserved, in certain rare cases, in cool localities of the mountains of central Europe or else disappeared entirely. The splits in range that thus arose are known as arcto-alpine disjunctions of range. They record the mutual relationships which existed in the past between the tundras and the treeless parts of the mountain ranges (HOLDHAUS 1929, 1954, EHRLICH 1958).

It was already recognised by DARWIN that the arcto-alpine type of distribution resulted from shifts of range during glacial periods. The arcto-alpine type of distribution is often confused with the boreo-alpine type displayed by central European forest species. Boreo-alpine species differ from arcto-alpine ones in their northern range, which includes the taiga zone but not the tundra. Arcto-alpine disjunctions arose during glacial periods, while the origin of boreo-alpine disjunctions is connected with the post-glacial period.

The Taiga Biome

This biome contains the largest uninterrupted regions of forest in the world. Apart from the tundras which lie adjacent to it on the north, it is the only biome which stretches right round the earth, interrupted only by the oceans. It forms a zone on average 1500 km wide. The environment in which the taiga fauna lives is characterised by cold climates in winter with temperatures down to −78°C (in East Siberia). The summers are short but warm with a growth period for vegetation of three to five months. There are extensive permafrost soils – in the northern part of the taiga – and widespread podsols and peat moors. Also characteristic are monotonous-looking conifer forests with relatively few types of tree (*Abies*, *Pinus*, *Larix*, *Picea*, *Salix*, *Betula*, *Alnus*, *Populus*).

The vertebrates of the taiga have many connections with those of the broad-leaved forest (Silvaea). Taiga species that occur both in the Old World and the New are: the lynx (*Lynx lynx*), the wolverine (*Gulo gulo*), the mink (*Mustela lutreola*), the stoat (*Mustela erminea*), Tengmalm's owl (*Aegolius funereus*) the three-toed woodpecker (*Picoides tridactyloides*) and the elk (*Alces alces*), although the latter is more associated with the widespread swamp and moor regions of the taiga. Characteristic species of the Eurasian taiga biome are: the nutcracker (*Nucifraga cariocatactes*), the waxwing (*Bombycilla garrulus*), the crossbill (*Loxia*) and the capercaillie (*Tetrao urogallus*). In accordance with the thin layer of humus that covers the ground in the conifer forests, the soil fauna consists of only a few groups such as mites, insect larvae, nematodes and apterygotous insects. Snails play only a small rôle. They are lacking in many conifer forests. On the other hand the wood wasps, sawflies and gall flies and the ants (*Formica*, *Camponotus*) are characteristic forms among the insects. Numerous forest pests are adapted to the conifer-forest belt; they include the spruce beetle (*Ips typographicus*), the pine looper moth (*Bupalus piniarius*) and the pine beauty (*Panolis piniperda*). The autumnal moth (*Oporina autumnata*), together with reindeer and man, is the most important ecological factor affecting the limit of the birch forests in Scandinavia (Tenow 1972). Within the insect fauna flightless forms occur, as in the oreal and tundra biomes. 'Wing reduction, which may have a variety of causes, is common in insects living in low-temperature environments. In insects generally and crane flies in particular living in cold latitudes or at high elevations or occurring as adults in winter in temperate regions, reduction or loss of wings has probably resulted from the insects' inability to use them in flight, so that natural selection would not act unfavourably on mutant forms in which reduction occurred.' (BYERS 1969; further examples in UDVARDY 1969).

Fig. 74. Mixed forest at the contact between boreal coniferous forest and broad-leaved temperate forest or *silvaea* (Canada, 1972).

The Silvaea Biome

This major environment, the broad-leaved temperate forest, is characteristic for the temperate areas of eastern North America, east Asia and central Europe. It has greater ecological variety than the taiga. This results in niches for a more diverse fauna, though the latter has of course been decisively altered by man. Human activities have produced new habitats for small animals, and many species owe their occurrence in central Europe or North America exclusively to man. As against this, many vertebrates, – such as the lynx, wild cat, wolf, bear, otter and beaver – are becoming scarcer and scarcer or, as with the aurochs, have already been wiped out.

The ecological association of individual silvaea animals has been especially well studied in beech forests (eg. FUNKE 1971, 1972). ANT (1969) was able to divide up individual beech woods on the basis of a number of species of snails. Thus the occurrence of *Abida secale*, *Clausilia parvula* and *Cepaea hortensis* is coupled with beech wood conditions. THIELE (1964) was able to show similar relations with ground beetles. The biotopes of *Abax ater* and *Abax ovalis* point to forest conditions near ground level.

In a similar way the beetle fauna of many half-arid stretches of turf in the Moselle area predominantly consists of an original, warmth-loving forest fauna (MÜLLER 1971, BECKER 1972). But in addition there are species among the butterflies and vertebrates which can only live in open landscapes.

Brown earths are the characteristic soil type of the silvaea.

For the biomass above ground level of a 120 year-old west European silvaea biome DUVIGNEAUD (1962) gives the following values, dry weight, per hectare:

Trees:	leaves	4t
	branches	30t
	trunk wood	200t
Undergrowth		1t
	total	235t
Birds		1.3kg
Large mammals:		2.2kg
Small mammals:		5.0kg
	total	8.5kg

Fig. 75. The *silvaea* biome or broad-leaved temperate forest (Appalachians, U.S.A., 1972).

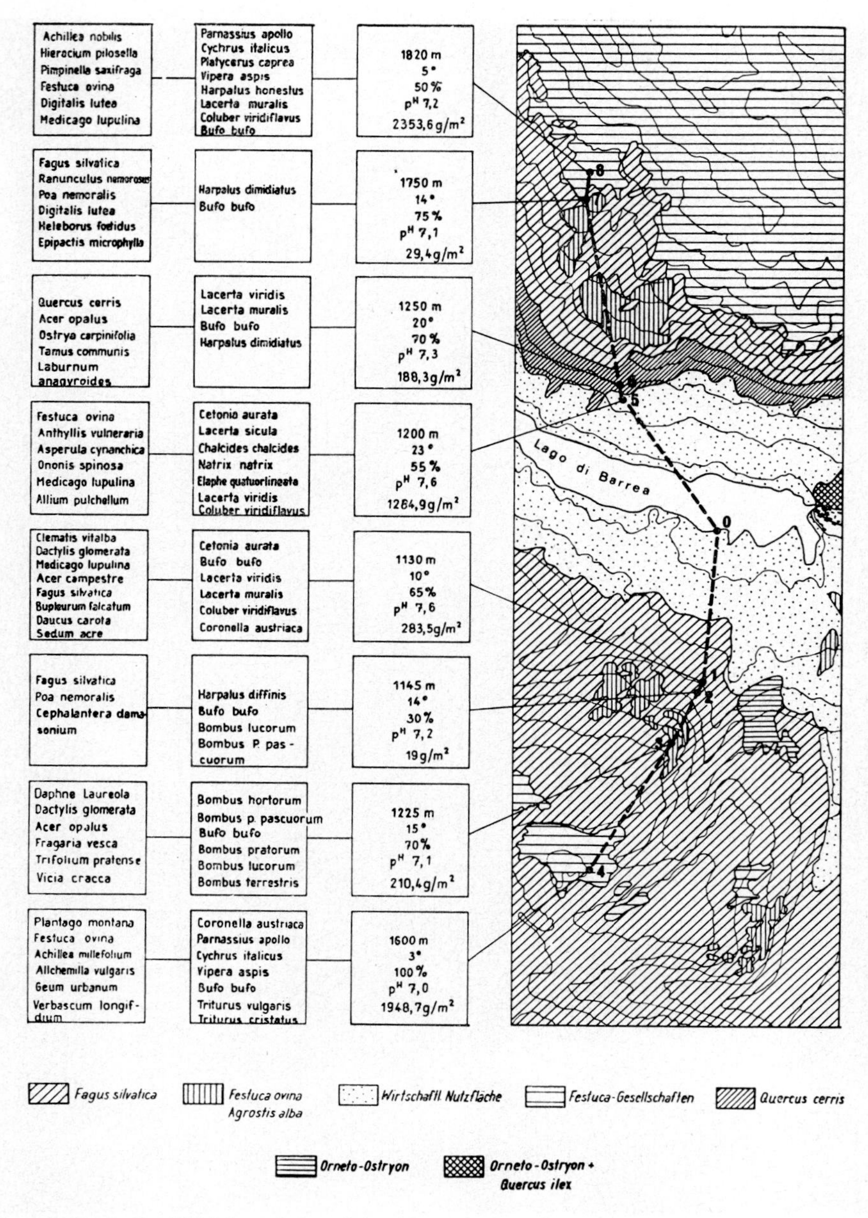

Fig. 76. Vertical subdivision of the *silvaea* biome in the Abruzzi National park (Italy).

In accordance with the ecological similarity of the disjunct silvaea biomes, there are many convergences and also strikingly close phylogenetic affinities between them. This is true both for the animals and for the plants. The moss flora of the North American Appalachians consists of more than 353 species. 196 of these, or 55.3%, also occur in Japan (IWATSUKI 1972).

Species of *Acer*, *Quercus* and *Fagus* are characteristic for the silvaea biome of the Nearctic region (*A. saccharinum*, *Q. rubra*, *Q. macrocarpa*, *F. grandiflora*) as well as for the western Palaearctic region (*Q. petraea*, *Q. robur*, *Fagus sylvatica* etc.).

A biome at the boundary between land and sea is the mangrove biome. In it lives a fauna restricted to the mangroves of the tidal zone of tropical regions. This fauna is poor in species but often extremely rich in individuals. It is often adapted to an amphibious mode of life and able to tolerate strong oscillations in salinity. It is able to incorporate in its food chains the abundant organic ooze which is deposited in mangrove conditions (ERICHSON 1923, FABER 1923, HENCKEL 1963, HUBERMANN 1959, SCHOLANDER 1968). Worthy of mention are the numerous crabs of the mangroves (*Uca*, *Sesarma*, *Cardisoma* etc.), the Gobiid fish genera *Periophthalmus* and *Boleophthalmus*, which are adapted to life in both air and water and the tree oysters (*Ostrea arborea*) which attach themselves to the roots of the *Rhizophora* mangrove. The fossil shells of these oysters can be used to date displacements in the geographical range of the mangroves during the Pleistocene.

Fig. 77. The mangrove biome (near Mombasa, Kenya, 1972).

Many smaller **natural** and **man-made ecosystems** are incorporated in the biomes. These smaller systems appear to have their own dynamics, but are nevertheless bound into the surrounding larger systems in many different ways. Three examples scattered through the land biomes are rivers, lakes and towns.

Fig. 78. Man-made Garigue in the Alpilles (Southern France, 1973).

Fig. 79. Man-made steppe landscape in the Grau (Southern France, 1973).

Fig. 80. Man-made *Pinus pinea* ecosystem (Algaida, near Sanlucar de Barrameda, Southern Spain 1973).

Fig. 81. Man-made ecosystem, Tea landscape in Ceylon (Pusselawa, 4.4.1974).

Fig. 82. Man-made ecosystem, Rice fields in Ceylon (Ambepussa, 8.4. 1974).

RIVERS AND STREAMS

Rivers and streams vary in speed of flow, temperature, suspended material etc. They are characterised by a fauna and flora adapted to these environmental conditions as they alter in the different parts of the course (ILLIES 1967, HUSMANN 1970, SCHWOERBEL 1971, ELLENBERG 1973, SIOLI 1954, 1955, 1968, STEFFAN 1972).

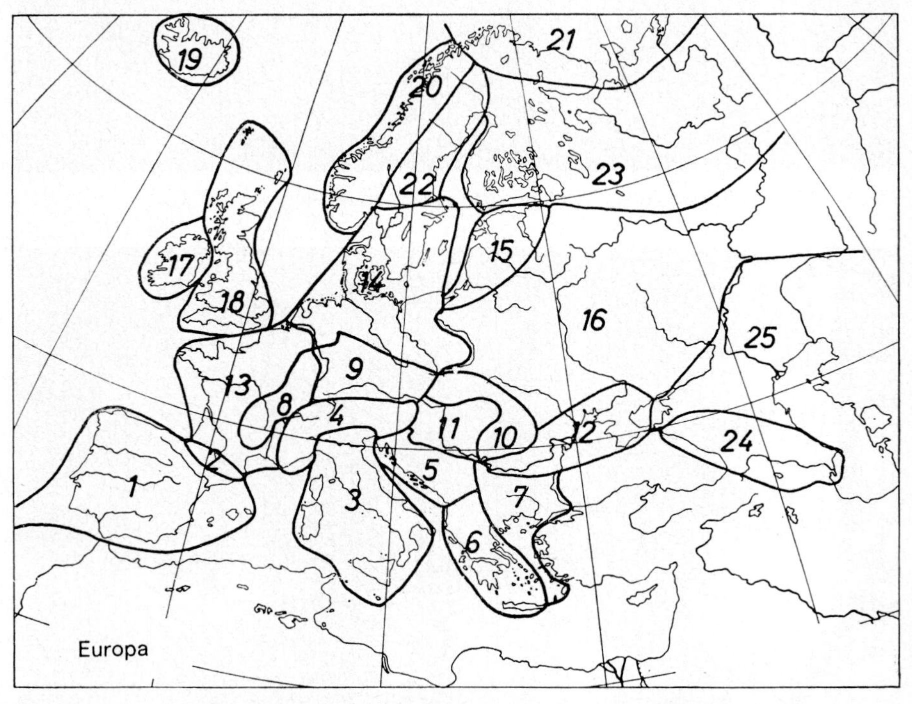

Fig. 83. Division of the European limnofauna into regions (after ILLIES 1967); The geographical regions: 1. Iberian peninsula, 2. Pyrenees, 3. Italy, 4. Alps, 5. Dinaric Western Balkans, 6. Hellenic Western Balkans, 7. Eastern Balkans, 8. Western highlands, 9. Central highlands, 10. Carpathians, 11. Hungarian Plain, 12. Pontic Province, 13. Western Lowland, 14. Central Lowland, 15. Baltic province, 16. Eastern Lowland, 17. Ireland, 18. England, 19. Iceland, 20. Boreal Highland, 21. Tundra, 22. North Sweden, 23. Taiga, 24. Caucasus, 25. Caspian Depression.

The upper reaches or rhitral have species adapted to high speeds of flow, whereas the species that occur in the lower reaches (potamal) are adapted to greater variations in temperature and slow currents.

At the places where they rise streams have a characteristic spring fauna made up of water animals and wet-loving land animals. This fauna occurs particularly in the ground water of the meadows round springs or accompanied by a spring flora of algae, mosses and higher seed plants. Water snails of the genus *Bithynella*, which are restricted to a temperature below 8°C, occur in such places. According to how the spring comes out of the ground it is possible to distinguish rheocrene springs, where the water immediately flows away; limnocrene springs, where the water flows first into a spring pool; and helocrene springs, where the water comes up in a bog.

The courses of central European streams and rivers can be divided into at least four regions on the basis of their characteristic fish fauna, seasonal oscillations in temperature and the nature of their beds. These four regions have their indicator species. The highest, brook region is the trout region, with *Trutta fario*. It is characterised by gravelly or rocky beds, and clear oligotrophic, oxygen-rich water with uniformly cool temperatures throughout the year. The trout region passes into the grayling region with *Thymallus thymallus*, which in general has warmer water and a partly sandy stream bed. After this comes the barbel region, with *Barbus barbus*, corresponding to the fast flowing reaches of rivers. The water in this region begins to be made turbid by material from the bed, which is sand mixed with mud. The barbel region passes into the bream region, with *Abramis brama*, corresponding to the lowland reaches with slow flow and muddy bottom. The species which occur in this region are adapted to turbid, warm water, poor in oxygen (for the Ecology of estuaries cf. McLusky 1971).

River and stream regions, after Schwoerbel 1971

		Corresponding fish zones
Crenal	Spring zone	Spring fauna
Rhitral	Mountain stream zone	Salmonid region
Epirhitral	Upper mountain stream zone	Upper trout region
Metarhitral	Middle mountain stream zone	Lower trout region
Hyporhitral	Lower mountain stream zone	Grayling region
Potamal	Lowland river zone	
Epipotamal	Upper lowland river zone	Barbel region
Metapotamal	Middle lowland river zone	Bream region
Hypopotamal	Lower lowland river zone	Pope-flounder region

Comparative studies have shown that the extent of the rhitral or of the potamal zone depends on altitude and latitude. At a given altitude the area of the rhitral zone decreases and that of the potamal zone increases in passing from the polar regions to the tropics (Schwoerbel 1971). The same displacement of zones is seen in a mountain range on passing from the higher regions down to the plain.

A corresponding pattern of distribution can be shown in brooks and rivers for other types of animal such as the flatworms or planarians. Thus *Planaria alpina* is characteristic for the upper reaches of running waters, *Polycelis cornuta* for the middle reaches and *Planaria gonocephala* for the lower reaches.

Human influences on running waters, especially the dumping of industrial sewage in them, can obliterate the natural biota of particular regions of rivers, with their characteristic species of plants and animals, or else can wipe out the boundaries between regions. New species appear, including saprobes, and indicate the change in the quality of the river by their presence and spatial distribution (c.f. WILLIAMS 1972).

The change of the Macrobenthos for the upper- and middle Rhine from 1917 to 1971 (after KINZELBACH 1972).

	1917	1957	1971
Porifera	5	2	1
Coelenterata	5	—	1
Turbellaria	?	?	2
Annelida	7	11	5
Crustacea	3	3	5
Insecta	39	10	3
Mollusca	20	14	9
Bryozoa	3	1	—
Total	82	41	26

A grievous disturbance of the biocoenosis is caused by the draining of indestructible substances into rivers. As well as sewage, these include agricultural pesticides to an ever-increasing extent.

Three groups of such substances can be distinguished, according to KÜHNELT (1970): 1) Concentrative poisons, whose effect is proportional to the dose. 2) Cumulative poisons. These have no noticeable effect in low concentrations. However they are stored within the organisms and show their effects when they reach a particular concentration, or when they are mobilised by a metabolic change. Examples are chlorinated hydrocarbons, organic phosphorus compounds, dinitro compounds and organic compounds of heavy metals, among others. 3) Summative poisons. These are destroyed in the organism or excreted again. However even in low concentration they cause irreversible damage to tissue. Examples are nicotine and cumarin derivatives.

A chlorinated hydrocarbon whose toxicity has been particularly well studied is endosulphane. When investigating the recent Rhine catastrophe GREVE and VERSCHUREN (1971) showed that endosulphane was carried adsorbed to particles of bottom mud. Laboratory studies of its toxicity to fishes showed that the LC_{50} value – i.e. the concentration of poison at which 50% of the experimental animals survived a given time (BAUER 1961) -- depends on the temperature of the water. A slight increase in the concentration of endosulphate causes a sharp rise in the toxicity (HERBST 1971).

However near the inflow of warm water into rivers, as from power stations, there is not only an increased danger due to faster chemical processes. In warmer water, warmth-loving species appear which cannot survive in normal temperate rivers. An example of this is the 'tropical' Tubificid worm *Branchiura sowerbyi* (TOBIAS 1972).

In addition to the immediate effect of increased temperature and decreased oxygen content on the fishes, the annual life-cycle is also disturbed. In extreme cases hibernation no longer occurs, and, because of the lack of available food in winter, the stock sometimes starve to death. A change in the spawning time, and quicker egg development because of higher temperatures also involve the danger that the young will hatch during the period of reduced availability of food. Still more important is the effect of increased temperature on the de-

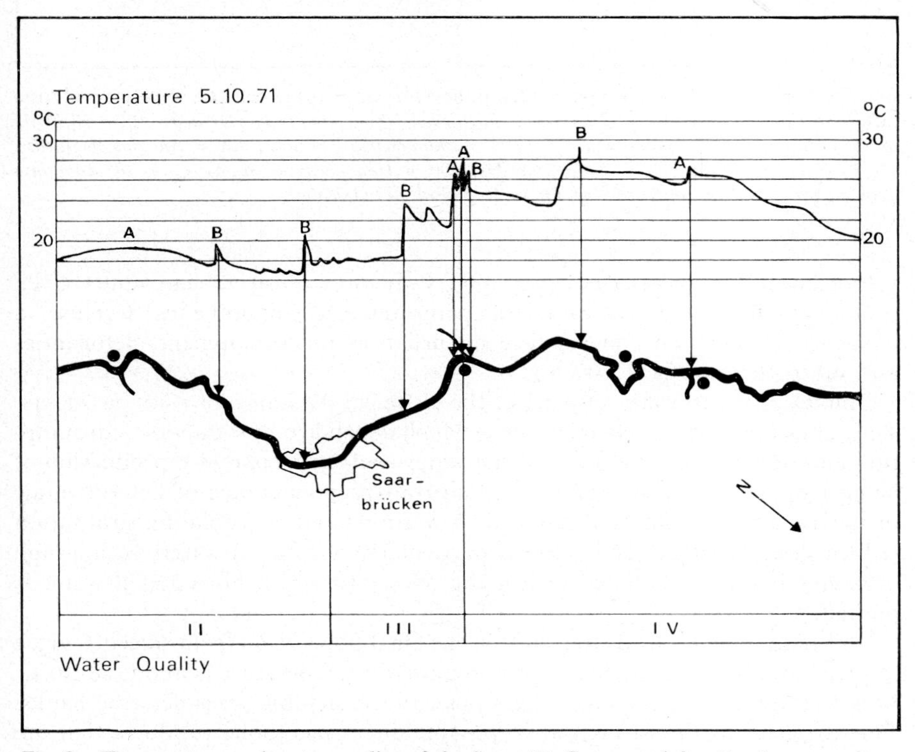

Fig. 84. Temperature and water quality of the Saar, W. Germany (after SCHÄFER, 1974).

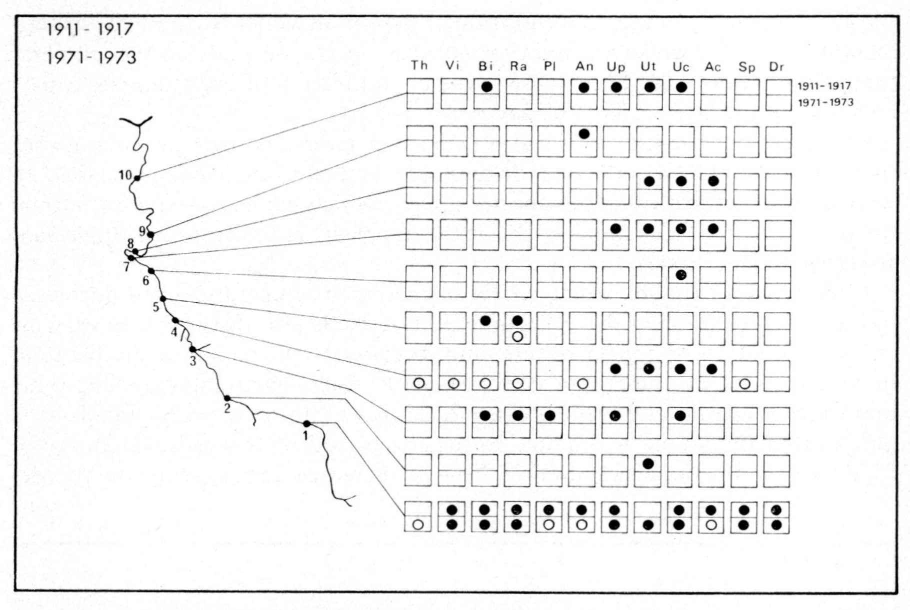

Fig. 85. Species that have disappeared from the Saar, since 1911, because of water pollution (after SCHÄFER 1974); Th = *Theodoxus fluviatilis*, Vi = *Viviparus viviparus*, Bi = *Bithynia tentaculata*, Re = *Radix peregra ovata*, Pl = *Planorbarius corneus*, An = *Ancylus fluviatilis*, Up = *Unio pictorum*, Ut = *Unio tumidus*, Uc = *Unio crassus batavus*, Ac = *Anodonta cygnea*, Sp = *Sphaerium corneum*, Dr = *Dreissena polymorpha*.

velopment of the spawn, since eggs and fry are more sensitive than adults. Even if the eggs do not die at raised temperatures, the simultaneous decrease in oxygen concentration can produce disturbances in development, deformities and subsequent losses (LAWA 1971).

Inflows of warm water also affect the chemical dynamics of running waters. They alter the biogenic chemical cycle which depends on production, consumption and destruction. Warming of the water leads to increased reproduction of heterotrophic organisms and thus to an over-preponderance of heterotrophic processes. The O_2 content is reduced by warming and microbial mineralisation reduces it still further. This effect is particularly obvious in waters with a high organic content, and these are just the ones into which hot coolant-water is poured.

Warming of water leads to a shift in the algal population from siliceous algae to green and blue algae whose optimum growth temperature is about 28–30°C. Massive flowering of the algae takes place in the shallow areas near the banks. This can lead to the formation of phytotoxins – metabolic products that are poisonous to warm-blooded animals.

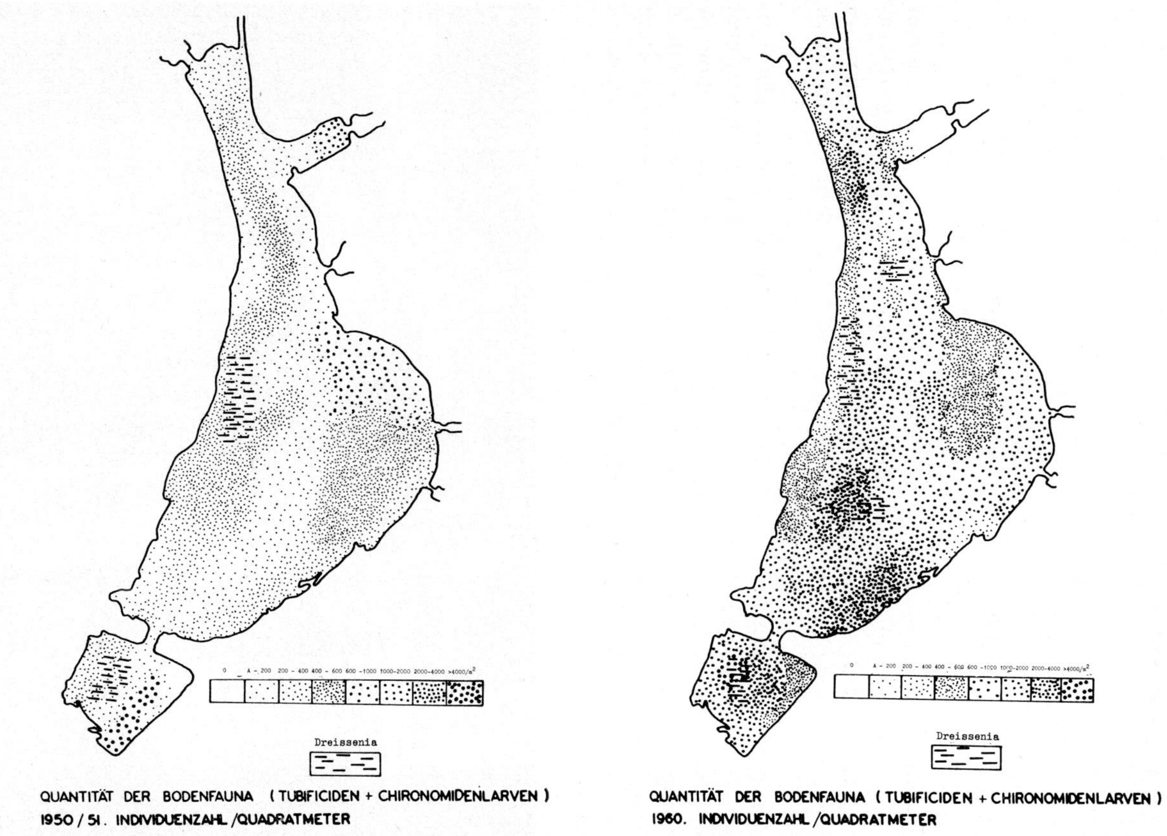

Fig. 86. Development of *Tubificid* and *Chironomid* populations in the Alster at Hamburg from 1950 to 1960, as indicators of increasing eutrophication (after CASPERS & MANN 1960).

LAKES AND PONDS

Biogeography of lakes and ponds is concerned with the study of the structure of aquatic ecosystems and functional processes in the freshwater habitat (OVERBECK 1972). Lakes be divided, according to their biogeographical peculiarities into polar, temperate and tropical lakes. Polar lakes have their surface frozen for several months in the year and their warmest water at depth.

The uppermost water layers of lakes generally have sufficient oxygen (epilimnion) and show a uniform temperature throughout because of the continual stirring that they undergo. Beneath the epilimnion is the metalimnion or thermocline in which the temperature quickly decreases going downwards until it falls to the temperature of the deepest layer or hypolimnion. The temperature of the hypolimnion approaches the value of 4°C, at which water has its greatest specific weight.

Fig. 87. Tropical lake with *Victoria regia* (near Manaus 1964).

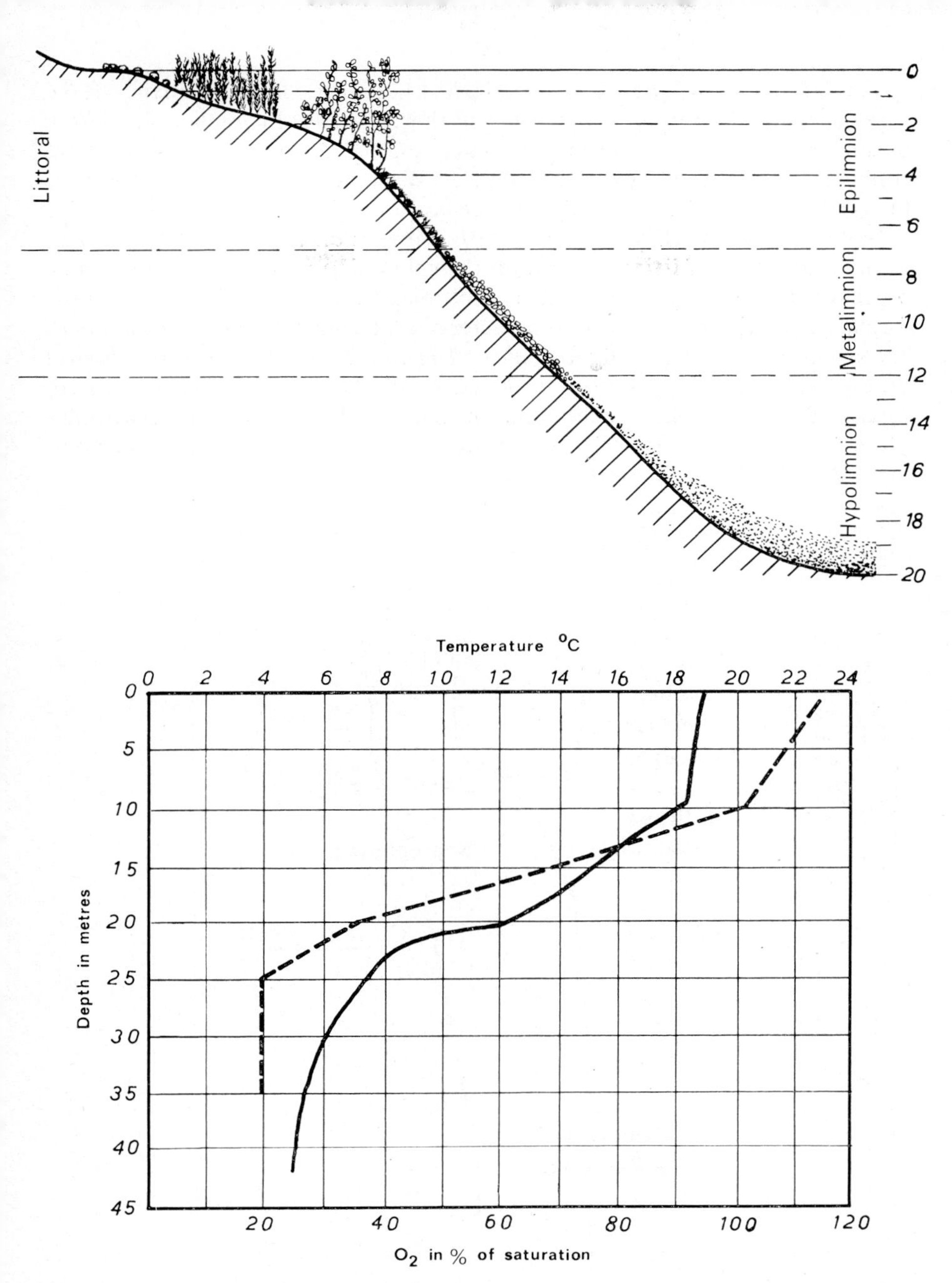

Fig. 88. Typical vertical distribution of temperature and vegetation in a central European lake (after DE LATTIN 1967).

This stratification is destroyed in so-called holomictic lakes by mixing of the superficial and deep layer of water during short periods of complete circulation in spring and autumn. Meromictic lakes, on the other hand, do not have complete mixing. In many cases, as with the Black Sea, this is due to differences in specific weight of the individual water layers.

Biological productivity can vary greatly between different lakes. An increase in the proportion of living material in the epilimnion may lead to more active breakdown by the bottom organisms. This may lead to lack of oxygen and the formation of the organic ooze characteristic of eutrophic lakes (*Chironimus* lakes). On the other hand oligotrophic or *Tanytarsus* lakes have lower biological productivity, linked with lower production of organic ooze and a sufficient supply of oxygen in the hypolimnion. In shallow lakes the danger of eutrophication through the inflow of sewage water is particularly high and can lead to the 'turning-over' of the water body.

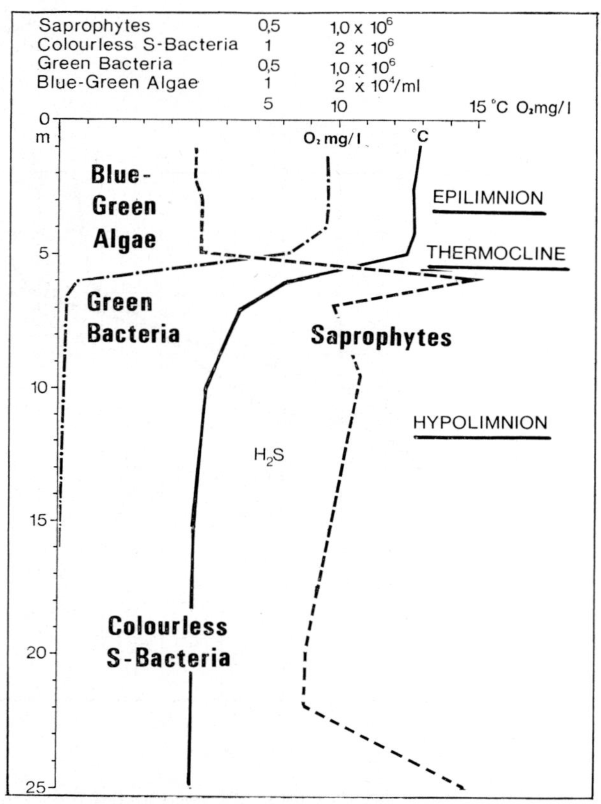

Fig. 89. Stratification from the Plußsee (Holstein, W. Germany; after OVERBECK 1972).

The Bodensee (Lake Constance) used to be a typical *Tanytarsus* lake. The most striking changes in the last few decades are the increase in its phosphate content, a raising of the level of plankton production and an increased consumption of oxygen in and below the thermocline, caused by the breakdown of this mass of plankton. Studies, using radioactive carbon, on the intensity of production of organic material by plant plankton shows that maximum production in bright weather is usually at a depth of 1 to 2 m, while the greatest proportion of organic material is produced in the top 5 m of the lake. The total production per unit surface area exceeds the values for oligotrophic lakes, and approaches that of eutrophic lakes. The winter production of organic material is limited by the circulation of the water masses, but in the winter of 1963, for example, it reached, beneath the ice, almost to summer values. The total annual production of living substance in the Bodensee amounts to about 2 million tons. Complete breakdown of this would use up about 130,000 tons of oxygen (ELSTER, Bodensee Project, 1968).

The Bodensee is at present in a labile transitional state between an oliogotrophic type of lake, low in organic production and low in nutrients, and a eutrophic type of lake, high in production and high in nutrients. Its increasing dirtiness is due to overloading with unpurified or only partly purified sewage. The outflows from households and from agricultural and industrial enterprises carry phosphatic and nitrogenous compounds into it. These cause an excessive growth of algae, and over-fertilize the water. When the algae die oxygen is required for their breakdown and this is extracted from the water. This leads to a gradually increasing poverty in oxygen. But when there is a shortfall of oxygen, biological self-purification is restricted. The lake suddenly changes from a clear, well aerated, oxygen-rich body of water into a sort of fluid overgrown with algae. Breakdown processes in the water then take place anaerobically, without oxygen. Poisonous substances are formed such as hydrogen sulphide, ammonia and methane. Of a total of 36,000 tons per annum of oxygen-consuming organic substances, 21% comes from the villages and town round the edge of the lakes; and these communities alone contribute 9% of a total of 17,900 tons of nitrogen and 20% of a total of 1,750 tons of phosphorous. Runoff of fertilisers from agricultural areas contributes 12,640 tons of nitrogen, or 74% of the total nitrogen content of the inflow, and 650 tons of phosphorus or 37% of the phosphorus content of the inflow. Out of a total contribution of phosphorus of 1,750 tons, 63% comes from the sewers of villages and towns on the lake's edge or from inflowing rivers. Comparing these figures with the quality of the Bodensee it is easy to understand the poor quality of the water where rivers flow into it or near industrial sites and towns (WAGNER 1967, MÜLLER 1966, LANG 1969).

Studies of the saturated pore-spaces in the sand of the margins of the Bodensee reveal a further degradation of the physical, chemical and biological situation in the eutrophic parts of the lake edge.

Average ammonium, nitrate and phosphate concentrations of rivers flowing into the Bodensee

	Flow m^3/sec	Ammonium N.	Nitrate N.	Phosphate P.
Neuer and Alter Rhein, Bregenzer Ach and Argen	313	3.8 t/day	14 t/day	0.32 t/day
		140mg/m^3	520 mg/m^3	12 mg/m^3
Other affluents	30	1.6 t/day	2.3 t/day	0.39 t/day
		600 mg/m^3	900 mg/m^3	150 mg/m^3
All affluents	343	5.4 t/day	16 t/day	0.71 t/day
		180 mg/m^3	550 mg/m^3	24 mg/m^3

The habitats of the mesopsammon of the Bodensee are almost entirely lacking in turbellaria. This corresponds to the amount of oxygen which the various species of turbellarians require. The density in individuals and the species make-up of the Tubificid populations depend on the depth of water, the supply of sediment and the type and specific action of the material undergoing sedimentation. In the lake-edge zone, down to about 30 m depth, species of *Euilyodrilus* and *Limnodrilus* predominate. In depths of 60 to 90 m the species *Tubifex tubifex* is dominant. With increasing sedimentation of decomposable organic material the maximum population density is displaced downwards. The species of *Euilyodrilus* and *Limnodrilus* move about 40 m deeper, and *Tubifex tubifex* about 100 m deeper than their appropriate depth zone in the Bodensee. In sedimentation areas where domestic sewage preponderates, species of *Euilyodrilus* and *Limnodrilus* are of first importance. With the inflow of industrial sewage *Tubifex tubifex* becomes dominant.

Bodies of standing water can be classified according to their content of suspended and dissolved material, especially humic acids. Dystrophic or brown-water lakes such as moor-lakes, with high humus content, have a shallow depth of visibility. They are contrasted with clear-water lakes with low turbidity. Bodies of standing water seldom persist through a long geological time span. Exceptions are Lakes Baikal, Tanganyika, Nyassa, Victoria, Titicaca and Ochrid. Of these Lake Baikal has existed since the Cretaceous as a separate

The number of cichlids and their degree of endemism in several of the larger and older lakes in east and central Africe is also very high (cf. GREENWOOD 1973).

Lake	Cichlids				Non-Cichlids			
	Species		Genera		Species		Genera	
	Total	Endemic	Total	Endemic	Total	Endemic	Total	Endemic
Victoria	ca 170	ca 164	8	4	38	17	20	1
Tanganyika	126	126	37	33	67	47	29	7
Malawi	ca 200	ca 196	23	20	ca 44	a 28	19	0
Albert	10	5	3	0	36	3	21	0
Rudolf	ca 7	ca 4	3	0	32	4	22	0
Edward/George	ca 40	ca 38	4	0	20	2?	10	0
Nabugabo	10	5	4	0	14	0	11	0

fresh-water sea and even remained open during glacial periods. This has led to the evolution of numerous endemics, including the sponge family Lubornirskiidae, the fish families Comephoridae and Cottocomephoridae and the snail families Baikaliidae and Benedictiidae. According to KOZHOV (1963), out of 652 animal groups that occur in Lake Baikal, 583 are endemic.
Like islands or isolated caves, isolate lakes are natural field experiments that offer an insight into phylogenetic phenomena. Lake Lanao in the Philippines and the East African lakes with their species-rich fish fauna have provided important information in this respect. The phylogenetic affinities of the limnofauna, in accordance with the ecological association of limnia organisms, have led to proposals for regional zoogeographical subdivision (ILLIES 1967) which often differ from those of the land fauna (BANARESCU 1967, 1970).

The postglacial evolution of the Baltic Sea has been characterized by alternating fresh-water and salt-water conditions. These started in the early postglacial with the Baltic ice lake. This formed from meltwater south of the Scandinavian ice cap and increased in size with the northward retreat of the ice until it was as large as the present-day Baltic Sea. Between 14000 and 8000 BC similar marginal ice lakes formed south of the ice caps over the whole of the Holarctic realm (THIENEMANN 1928, 1950, SEGERSTRALE 1957, etc).

Colonization of the North American marginal ice lakes took place partly from eastern Siberia. North American ice lakes extended from the St. Lawrence river to the foothills of the Rocky Mountains in Saskatchewan.

Further retreat of the southern margin of the ice in the Baltic area opened a pass by way of Lakes Väner, Vätter and Mälar. Through this pass the fresh water flowed out into the sea. As a result the surface of the lake fell to sea level and also a direct marine connection through the Central Swedish lakes came into existence. In this way the Baltic ice lake was transformed for the duration of the period from 7500 to 7000 BC into an arm of the sea. This is termed the Yoldia sea after its index fossil, the bivalve *Yoldia arctica* (now called *Portlandia arctica*). Because of the continued isostatic elevation of Scandinavia the marine connection was once again interrupted and a new fresh-water lake came into existence. This was the Ancylus lake, named after its index fossil, the freshwater limpet *Ancylus fluviatilis*; the lake lasted from 7000 to 5000 BC. Continuous subsidence of the land between Jutland and southern Sweden, in the course of which the Belts and the Öresund broke through, produced the connection with the North Sea which still exists today, and the Baltic Sea once more received salt water. The salinity was at first higher than at present and this stage of the Baltic Sea, coinciding with the post-glacial climatic optimum, is known as the Littorina Sea, after its index fossil the periwinkle, *Littorina littorea*. At the present day *Littorina littorea* is confined to the western part of the Baltic Sea while fresh-water animals are again penetrating into the brackish water of the central and eastern parts of the Baltic Sea. The present-day stage of the Baltic Sea is named the Lymnaea Sea after the snail *Radix peregra* (*Lymnaea ovata*) which has migrated into it from fresh water (SEGERSTRALE 1954).

URBAN ECOSYSTEMS

An industrial town also represents an ecosystem (Southwick 1972, Müller 1973). In it, man is the critical key-species. Such biocoenoses, with mosaic distribution, very rarely allow much to be said concerning their historical development. The ways in which they normally react to changes of environmental conditions simply have to be used as indicators. The carrying capacity of town ecosystems depends on the group-specific behaviour of their inhabitants and the combined effect of all the relevant spatial factors. The limits of the carrying capacity of certain industrial towns have already been reached or exceeded, as with the Donora catastrophe of 1948. Sometimes they can only be worked out approximately (Bach 1972, Cralley 1972, Hill 1972, Lamb 1968, McBoyle 1972, Probald 1972, Turner 1964, Müller 1973 etc). It can be shown that economics and ecology are interdependent factors in the ecosystem of an industrial town. The working-out of a systematics of urban ecosystems has only begun very recently (Müller 1973, Ellenberg 1973).

The high concentration of emissions in urban air produces a canopy of fumes which determines the radiation economy of the town by decreasing the effective quantities of radiation received and given out. A decrease in effective outward radiation and the production of heat by other methods leads to a rise in temperature in the town as compared with the unbuilt-over countryside (cf. Bach 1972, Dansereau 1970, 1971, Hesketh 1973, Hill 1972, Lamb 1969, McBoyle 1972, Müller 1972, 1973, Probald 1972).

Although air humidity is lower in towns the formation of fogs and clouds increases because of a rise in the number of condensation nuclei. This leads to increased precipitation. Water cannot be stored because of the speed with which it runs away. Towns are drier than their surroundings. The urban ecosystem can be defined by a number of abiotic and biotic factors, including man.

Gases and solids are differently distributed in urban systems than in more natural ecosystems. This is particularly true of the concentration of SO_2 and CO, though it needs to be pointed out that the oceans, rather than man, have been the main source of CO (Eberan-Eberhorst 1972). The origin of air pollutants in urban ecosystems varies considerably with the region.

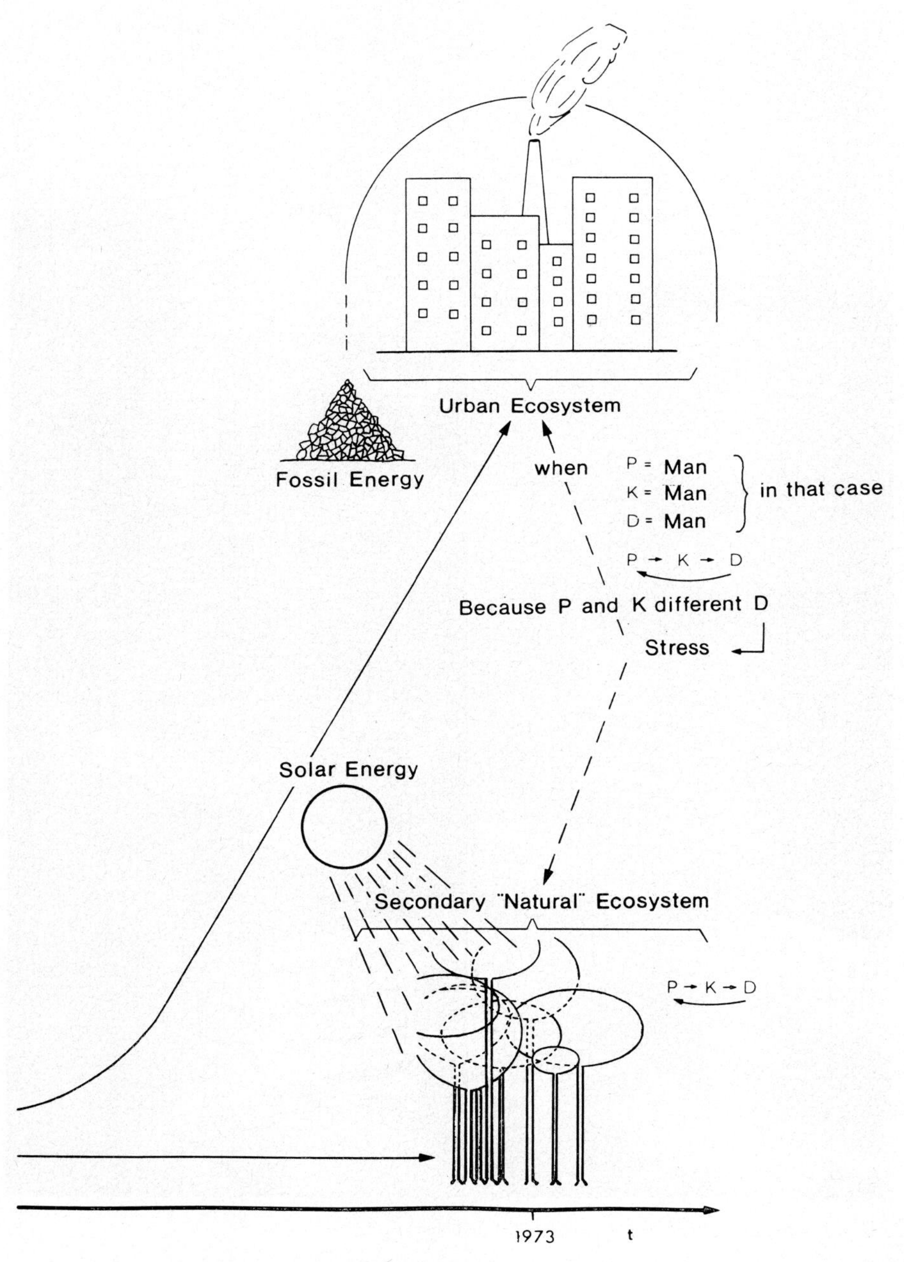

Fig. 90. Tropical rain forest (below) and industrial town (above). The two most extreme terrestrial ecocystems of the present day.

Fig. 91. Chicago as urban ecosystem (August 1972).

Fig. 92. *Biston betularia*, an example of industrial melanism; Left: melanic form; right: normal form.

Sources of air pollution in the U.S.A. (absolute values in 10^6 tons/year). Absolute and relative values after EBERAN-EBERHORST 1972.

	Solids	SO	NO	CO	HC	Total	Total %
Transport	1.8	0.5	3.1	59.6	9.7	74.7	60.7
Industry	6.0	8.7	1.6	1.8	3.7	21.8	17.7
Power stations	2.4	10.2	2.4	0.5	0.1	15.6	12.7
Space heating	1.2	3.4	0.8	1.8	0.5	7.7	6.3
Burning of rubbish	0.6	0.2	0.1	1.3	1.0	3.2	2.6
Total	12.5	23.0	8.0	65.0	15.0	123.0	100.0

Biocoenoses and industrial species distributed in a mosaic pattern permit towns to be defined very precisely from the floristic and faunistic viewpoints. Thus the dipterous fauna of a central European city has a characteristic make-up (PETERS 1949); species of *Fannia*, *Musca* and *Drosophila* are dominant (see KÜHNELT 1970 for further examples).

Two tendencies are obvious in the urban fauna: 1) there is a sudden decrease in the individual abundance and specific diversity of whole orders, 2) Particular species show a preference for the urban ecosystem.

Reptiles, amphibians and shelled snails are mostly quite absent or else very rare. Among grasshoppers only *Stenobothrus bicolor* has succeeded in penetrating into the central areas of towns. The aerial plankton in the centres of towns is differently constituted to what it is near the edges. Tardigrades seem to be lacking entirely in the middle of towns although their cysts can withstand extreme environmental conditions. On the other hand some species reach their optimum of population near town centres. This is true in summer of the swift (*Apus apus*). And in winter its place is taken by over-wintering starlings (*Sturnus vulgaris*) and collared doves (*Streptopelia decaocto*).

The fauna of houses has a special position within the urban fauna and mainly consists of exotic forms that require warmth. The cockroach *Periplaneta americana* has become a house animal particularly in university towns. The Indian pharaoh ant (*Monomorium pharaonis*) occurs in several European capitals. The North American termite *Reticulitermes flavipes* was introduced into Hamburg in 1937 where it has established itself in the municipal heating system.

Historical zoogeography can only begin when the various phases of zoogeography briefly discussed above have been successfully completed, when the dynamics of the organisms living together at the moment on earth has been understood and when the ecological connections of species, their population genetics (SPERLICH 1973) and their spatial and temporal distribution are known. Without these foundations mistaken conclusions are inevitable.

THE EVOLUTION OF THE DISTRIBUTION AREA

Historical zoogeography attempts to throw light on the historical evolution of areas of distribution. This task is only valid if an unbiassed attempt has been made to explain the origin of the landscapes which control the areas of distribution. Besides ranges which have remained constant for a very long time, there are others which show an extraordinary dynamism persisting up to the present. The age of the occurrence of a species at a particular locality is very seldom the same as the geological age of this locality.

The fact that a species or genus now occurs, say, in South America is far from indicating that it arose there. Precise historical analysis of a range presupposes knowledge of the facts that influence the range in its present-day environment and of the landscape dynamics and evolutionary dynamics of the species in the past. How this requirement can be satisfied will be discussed later.

If the occurrence of isolated groups of animals in various areas and countries were to be explained simply by the existence of former land or water bridges without regard to evolution and their present-day powers of dispersal, then there would be no place on earth that was not covered by some such bridge. It is important to notice that some of these bridges in fact existed, but they can in no way be proved by methods of this sort. It is precisely this 'bridge-building', often frivolous, which has done zoogeography so much damage.

Some of the bridges proposed will here be named and briefly discussed.

Schuchert Land was a North American-Pacific mountain range, which in the opinion of SCHUCHERT & IHERING (1927) connected North with South America during the Cretaceous. From west to east it stretched from the Pacific Coast to the Missouri Basin (part of the Tethys Sea).

Lemuria was a landbridge demanded by SCLATER and IHERING to connect Madagascar with India in the Cretaceous and early Tertiary. It was 'constructed' to explain the distribution, disjunct at the present day, of the Madagascan and Indian lemurs. More recent biogeographical studies have shown the independence of the modern Madagascan fauna (eg. GÜNTHER 1970) and have been able to demonstrate that much of it came in over the sea (PEAKE 1971).

Beringia was already suggested in BUFFON's time as a bridge between Eurasia and North America. More recent palaeontological, geological and biogeographical results indicate that Beringia existed several times during the Tertiary, and during the cold periods of the Pleistocene. An exchange of faunas took place over it which was very important for both the Old and New Worlds. Successive

Geological Periods and their ages in years after THENIUS (1972):

Era	Period	Central European divisions	Age in 10^2 years
	Quaternary	Holocene	0.01
		Pleistocene	2 – 2.5
		Pliocene	
Caenozoic	Tertiary		
		Miocene	25
		Oligocene	
		Eocene	
		Paleocene	65
		Senonian	
	Cretaceous	Turonian	
		Cenoman	
		Gault	
		Neocomian	135
Mesozoic	Jurassic	Malm	
		Dogger	
		Lias	180
	Triassic	Keuper	
		Muschelkalk	
		Buntsandstein	220–25
	Permian	Zechstein	
		Rotliegendes	280
Palaeozoic	Carboniferous	Upper Carboniferous	
		Lower Carboniferous	340–355
	Devonian	Upper Devonian	
		Middle Devonian	
		Lower Devonian	400
	Silurian		440
	Ordovician		500
	Cambrian		570–600
Precambrian	Algonkian		2000
	Archaean		>3000

marine and terrestrial deposits on the Bering-Chuktien platform with *Sequoia* etc. show that several land connections had often occurred already in early

Tertiary times. The land connection was broken at the beginning of the Quaternary by a marine transgression whose extent is marked by an old strand cliff stretching from the arctic coast of Alaska as far as the Yukon river. The snail genus *Neptunaea*, which in the Tertiary had been restricted to the Pacific, first appeared in Atlantic beds at this time. However, Beringia again came into existence through falls in sea level during the cold periods of the Pleistocene. But pollen analysis, coring data and C_{14} dating shows that the bridge was flooded in Interglacial periods. It finally disappeared at the beginning ot the postglacial climatic optimum, when man first entered America.
Archiplata, in IHERING's opinion (1927) ,was a landbridge in what is now Andean South America. It arose in the Cretaceous and connected the South Pacific landbridge (Archinotis) with Schuchert Land. A faunal exchange between North and South America took place by way of Archiplata during the Tertiary.
Archiguiana was a supposed Cretaceous island in the region of Venezuela and the Guayanas. The distribution of many genera known from only one species coincides with this island. Examples are the parrakeet *Gypopsitta vulturina*, the Cotingids *Haematoderus militaris* and *Perissocephalus tricolor* and the tyrant-flycatcher *Microcochlearis josephinae*. However later biogeographical studies show that these distributions arose through very recent displacements of the rain forest.
Mesozoic Archhelenis is supposed to have acted as a landbridge connecting South America with South West Africa by way of Tristan da Cunha. It was summoned up to explain the resemblances between the older faunas of Africa and South America. KOSSWIG (1944) was able to show that the affinities between African and South American animals could be satisfactorily explained in other ways. This is true for the Spirostrepids (millipedes, cf. KRAUS 1964), the clawed frogs, the Cichlid and Characid fishes, numerous parasites (*Nesoelecithus* cf. MANTER 1963), ostracods and Mutelids. It also holds for the species of *Mesosaurus*, only known as fossils, whose ranges include South America and also great parts of Africa (KURTEN 1967).
Archatlantis (IHERING 1927) was a supposed Cretaceous landbridge connecting the Antilles and Florida with North Africa and southern Spain and including the Azores, Canaries and Cape Verde islands. Archatlantis is not needed to explain the manatees (Trichedidae) occurring on both sides of the Atlantic. These animals are not known in pre-Pleistocene deposits.
Canaries bridge has recently been demanded on the basis of geological, palaeontological (SAUER & ROTHE 1972) and phylogenetic connections between the eastern Canary islands, including Lanzarote, and the African continental block; it is still under discussion whether there was actually a bridge, or, rather, a gradual separation of the islands one from another. 'Separation of the eastern Canaries from Africa might have been by rifting and a land connection might still have existed in the lower Pliocene' (SAUER & ROTHE 1972).
Tyrrhenian bridge was a supposed Riss- Glaciallandbridge between Toscany in

Italy and Corsica and Sardinia. It was demanded by biogeographers of the last century on the basis of the close phyletic affinities of the herpetofauna (cf. SCHNEIDER 1971). Other landbridges of various ages have also been supposed in this region so as to interpret the different faunal connections. Examples are: the **Galita bridge**, connecting Corsica and Sardinia with Africa; the **Balearic bridge** connecting Corsica and Sardinia and the Balearic islands with Spain; and the **Provence bridge**, connecting Corsica and Sardinia with Provence. Up to the present there have been no purposive studies on the ability of the animals of this region to be introduced passively.

The Pleistocene faunas of North Africa and Sicily, including dwarf elephants (cf. VAUFREY 1929), argue at least against a landbridge between Sicily and the African continent in glacial times. 'On the contrary, the fauna of Africa and the fauna of Italy are well individualised and very different, and the points of contact between them are too sporadic to be significant' (VAUFREY 1929).[1]

Archinotis was a landbridge that was already demanded by biogeographers of the past century for the Cretaceous and early Tertiary. It connected southern South America with New Zealand and Australia by way of Antarctica and the South Pacific islands.

In the nineteenth century HOOKER (1847) like RUETIMEYER (1867) required a land connection between Australia and South America by way of Antarctica. At the beginning of this century ORTMANN (1901), MEISENHEIMER (1904), OSBORN (1910) and HOFSTEN (1915) furnished further biogeographical evidence for the existence of a southern Pacific land connection. WITTMANN (1934) came on the basis of his studies, to the following summing-up: 'The Antarctic affinities become comprehensible as soon as one imagines a connection by way of an island bridge or a continental bridge. An island bridge, however, is not probable for ecological reasons, and a continental bridge is untenable for geophysical reasons. Consequently these antarctic affinities can only be explained on the supposition of large-scale warping.'[2]

This land connection is still demanded by biogeographers today, in order to explain in a sensible fashion the close affinities of the groups of organisms distributed on both sides of the southern Pacific (BRUNDIN 1966, ILLIES 1965, MERTENS 1972, MÜLLER & SCHMITHÜSEN 1970). This is true, for example, of the southern beech (*Nothofagus*), fresh-water crayfish of the family Parasticidae, stone flies of the family Eustheniidae, midges (Chironomidae), land snails (Bulimulidae) and snake-necked turtles (Chelidae) (cf. MONOD 1972, BRUNDIN 1972, HALFTER 1972, LAURENT 1972, GASKIN 1972, SCHMINKE 1972).

1 'Au contraire, faune d'Afrique et faune de Sicile sont bien individualisées et bien différentes, et les points de contact entre elles sont trop sporadiques pour être significatifs'.

2 'Die antarktischen Beziehungen werden verständlich, sobald man die Verknüpfung durch eine Inselbrücke oder einen Brückenkontinent herstellt. Die Inselbrücke ist aber aus ökologischen Gründen nicht wahrscheinlich, der Brückenkontinent aus geophysikalischen Gründen unhaltbar. Es sind also die antarktischen Beziehungen nur erklärbar bei Annahme einer Epeirophorese'.

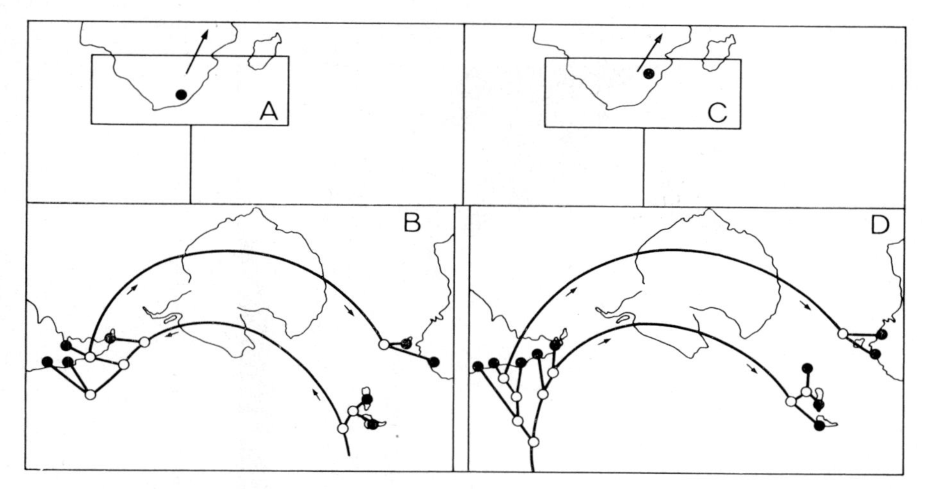

Fig. 93. Circum-Antarctic distribution and inferred transantarctic dispersal in the subfamilies Podonominae; A = *Boreochlini*, B = *Podonomini* and Diamesinae, C = *Diamesae*, D = *Heptagyiae*, Diptera *Chironomidae* (after BRUNDIN 1972).

Geologists have now collected decisive evidence to support the views of WEGENER (1929) concerning continental drift. For this reason **Gondwanaland** has again become important (cf. HALLAM 1973, REYMENT 1972, VANDEL 1972, GOSLINE 1972, CRACRAFT 1972, COLBERT 1972, PATTERSON 1972, PAULIAN 1972, AXELROD 1972, TARLING & RUNCORN 1973, PLUMSTEAD 1973). Gondwanaland was a Palaeozoic continent in the southern hemisphere which included South America, Africa, Madagascar, India and Australia and which, in WEGENER's opinion, was broken up by continental drift. Its connections in space are a key to explaining the distribution of many mesozoic and palaeozoic animals and plants. Examples are the Gondwana land flora with *Glossopteris* and the *Mesosaurus* group. North west of Gondwanaland lay the primeval continent of **Laurasia** including North America, Greenland, Scandinavia and parts of Siberia. According to WEGENER, Laurasia consisted in the Mesozoic of two continental blocks separated by the Tethyssea. One of these blocks, **Laurentia**, included the northern part of North America, while the other one, **Angaraland**, was the forerunner of Eurasia. The evolution of Peruvian and Mesozoic reptiles can be correlated with the supposed degree of separation of the continents during these periods. The turtles and their relations such as *Eunotosaurus*, and also the mesosaurs, eosuchians, rhynchocephalians and ornithischians evolved and diverged in Gondwanaland while the sauropterygians and therapsidans arose in Laurasia.

The significance of Laurasia and Gondwanaland for biogeographical interpretation was long controversial. NELSON (1969) pointed out that the value of historical biogeography depends on the 'repeatability' of its results. The geo-

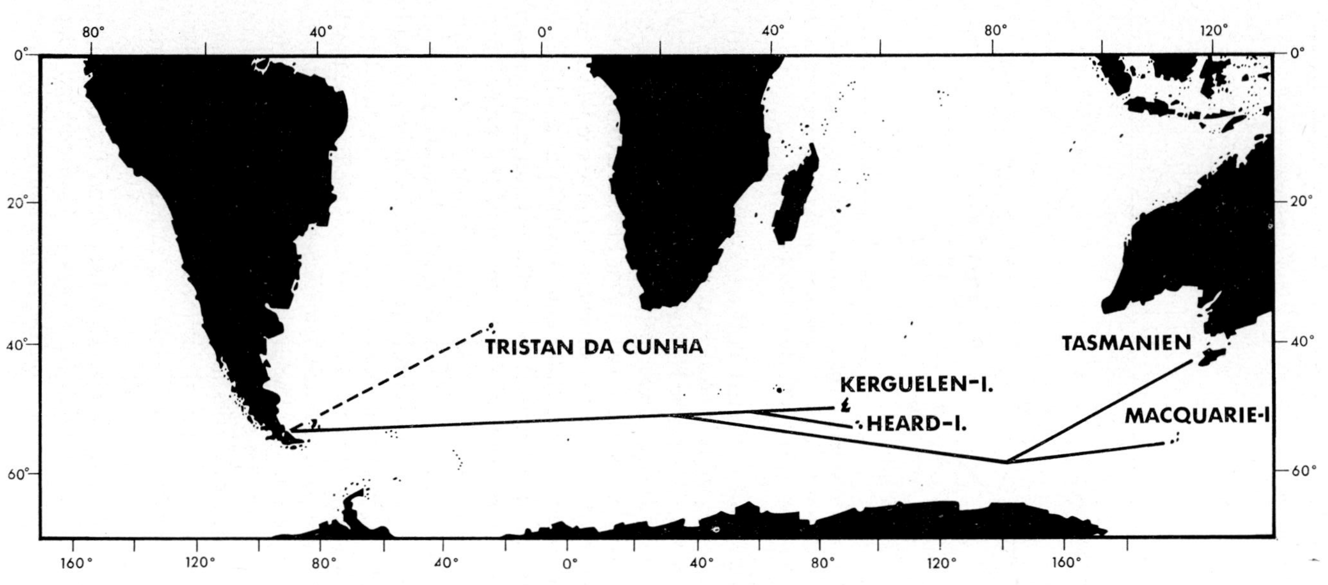

Fig. 94. Principal introduction routes for *Nothofagus* pollen and trees in the South Atlantic.

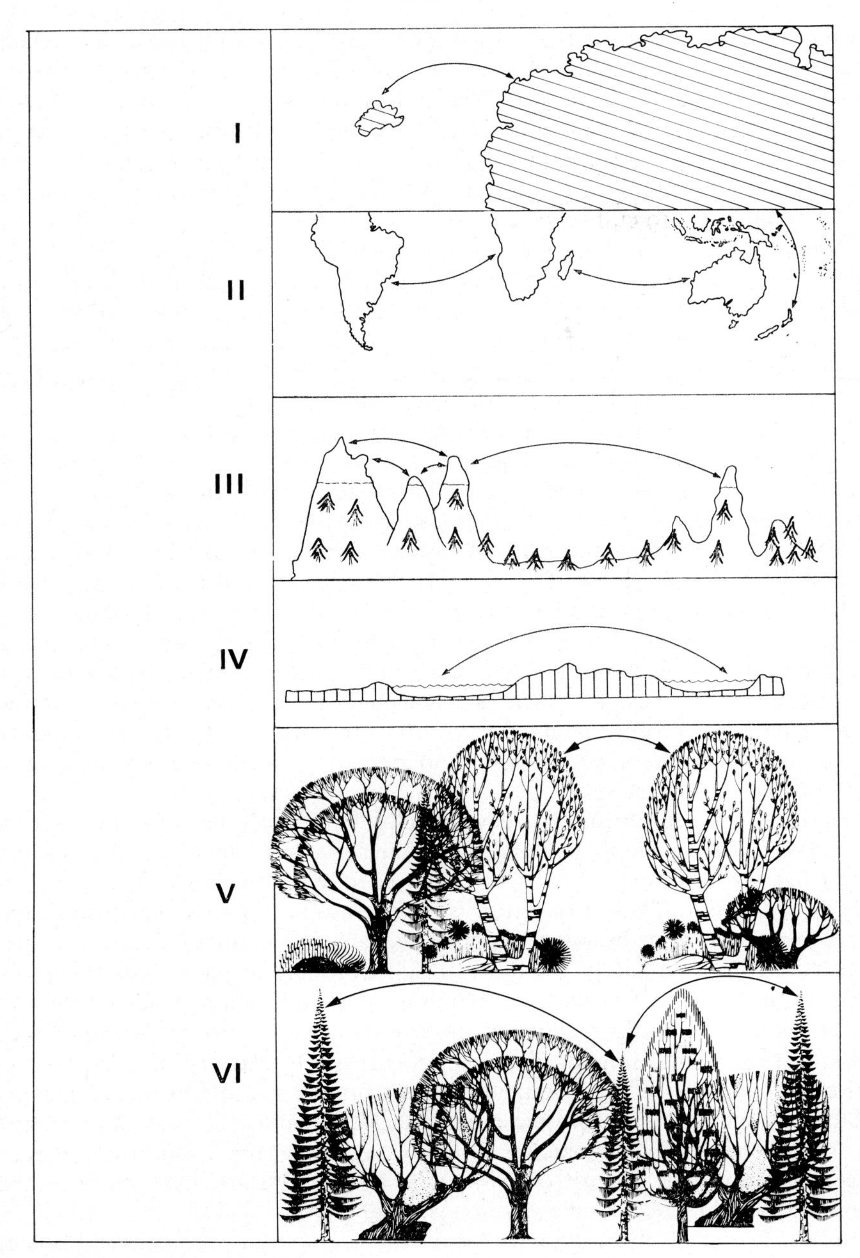

Fig. 95. Islands and Habitat-Islands.

physical results reached in the last few years have rested in part on an excellent zoogeographical foundation formed by the basic works of HENNIG (1960), BRUNDIN (1966) and ILLIES (1965). The phylogeny of species became in HENNIG's hands an indicator of the evolution of lands and landscapes. Critics, such as DARLINGTON (1970), have paid too little attention to the phylogenetic systematics of HENNING. Only this approach will have true significance for zoogeography (BRUNDIN 1972, PETERS 1972).

Geological events must be considered along with the dynamism of an animal's range, which is often extraordinarily large (cf. LINDSTRÖM 1970, CAPPETTA, RUSSELL & BRAILLON 1972, CHALONER & MEYEN 1973, KAUFFMAN 1973, PALMER 1973, TRALAU 1973). Species and individuals can drift all over the world. Only a precise study of the means of distribution available to a species makes that species significant for zoogeography.

More than any other type of study the investigation of islands has led to a deeper understanding of the possibilities of dispersal available to individual species. In working on **islands faunas**, biogeography can throw light on such basic phenomena as history of colonisation, dispersal, effect of competition, problems of adaptation and rates of supresession and extinction. All this provides important information for the study of evolution (CARLQUIST 1965, 1966). Coherent biogeographical problems can more easily be studied on islands. It is therefore no surprise that **island biogeography** has recently led to the introduction of a large number of quantitative methods and experimental estimates. Island faunas can be used as pointers in considering the dynamics of distribution of related species on continents. They throw light on facts of climatic and earth history and lead to a profounder understanding of present-day landscape relationships (MÜLLER 1969, 1972).

The theory of equilibrium was developed by PRESTON (1962) and MACARTHUR & WILSON (1963). These authors maintained that on islands an equilibrium exists between the number of species newly immigrating and the number of species dying out. This proposition is a very useful one even supposing that complete equilibrium between immigration-rate and extinction-rate will only be reached approximately. It allows predictions to be made and tested by experiment. The result of the first colonisation of an island depends on the size of the island and its distance from the source of colonisation. In accordance with this the colonisation-curve can vary greatly in the rate of climb. With time, however, it reaches a high point dependent on the size of the island and its ecological make-up. Further species can only be added if species already present die out. WILSON & SIMBERLOFF (1969) have tested the equilibrium theory experimentally. The theory of MACARTHUR and WILSON has provided the stimulus for many new studies (eg. NEVO et al. 1972).

In certain islands whose age of isolation and geological age can be established, the rate of colonisation can be studied.

A beautiful example of this is **Surtsey**. This is an island 2.7 km^2 in area situ-

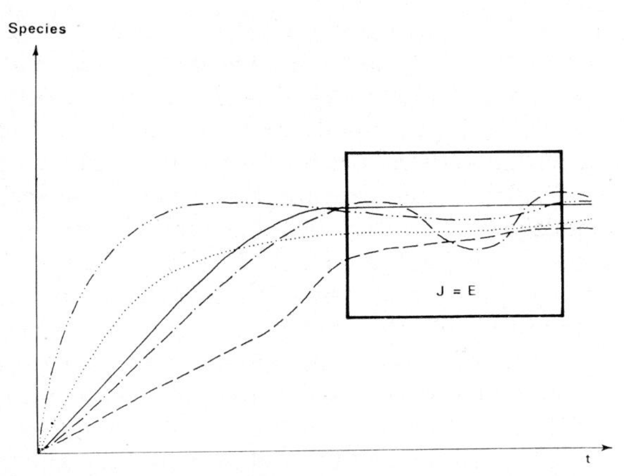

Fig. 96. The relationship between the number of species in a given surface area, and time. After a certain time the number of species remains constant. The rate of immigration corresponds to the extinction rate, producing equilibrium.

ated 30 km south of Iceland at 68°18′N, 20°36′W. It arose on 14th November 1963 as a result of a submarine volcanic eruption. The natural colonisation of the island, which was originally bare of life, has been studied since 1964 by many biologists and since 1965 has been worked on by the Surtsey Research Society (SCHWABE 1970, SCHWABE & BEHRE 1971). The research results of this society have been published since 1964 in Reykjavik in the Surtsey Research Progress Reports. The nearest island to Surtsey is the somewhat smaller island of Geirfuglasker 5 km away. The first insect was observed on Surtsey on 14th May 1964 and was the Chironomid *Diamesa zernyi.* In the autumn of 1964 a Noctuid moth, *Agrotis ypsilon,* was taken. Up to 1968 70 living arthropods could be collected on the island, among which diptera predominated with 43 species. Unlike the arthropods, which came to the island by air, the five species of higher plants whose existence had been established on Surtsey up to 1968 came by sea. Those were *Cakile edentula* and *Cakile maritima* (sea rockets), *Elymus arenarius* (lime grass), *Honkenya peploides* (sea purslane) and *Mertensia maritima* (northern shore-wort). No species of lichen had colonised Surtsey up to 1968. On the other hand several land algae had been found on the island. Surtsey is a fine example showing how a new, firmly-structured ecosystem can be built up on a sterile island as a result of chance.

A further example is **Krakatao** (cf. DAMMERMANN 1922, 1948). This is a volcanic island 832 m high lying 40 km west of Java. On the 26th to 27th August 1883 there was a volcanic eruption on the island. This reduced its area from 32.5 km^2 to 10.67 km^2 and destroyed all animal and plant life on it, although until then it had been entirely covered with forest up to its summit. In 1886 27 species of higher plants were proved to exist on the island, while in 1897 there were already 62 species and in 1906 a total of 114 species. Most of these came by air – the island of Sibesia is only 18.5 km distant. In 1889 40 species of arthropods, two reptiles and 16 birds occurred, while in 1923 there were about 500 arthropods, seven land gastropods, three reptiles including a snake, 26 species of breeding birds and three species of mammals comprising two bats and a rat. Surtsey and Krakatao illustrate by their make-up of species that organisms vary enormously in their dispersal abilities in particular stages of development. They also show that the ecological valency of a species is critical for its dispersal. Many species have the ability to disperse themselves, for example by flight, but make no use of this ability. Others, on the other hand, form a regularly occurring proportion of the aerial plankton. Above all, such islands show how other island faunas have arisen whose time relations are often difficult to elucidate (cf. GÜNTHER 1969). Two groups of islands lying outside the continental shelf will be mentioned here – Hawaii and the Galapagos.

The **Hawaiian islands** are about 16,680 km^2 in area. They have been seen by many scientists as an independent realm because their fauna is so rich in endemics. The group of islands is of Pliocene age and volcanic origin and has never had a land connection. The ancestors of the animals found there came in over the sea and developed into separate species. This is true for example of the bird family Drepaniidae or honeycreepers with eleven genera and 21 species and the snail families Achatinellidae and Amastridae. ZIMMERMAN (1948) has given a detailed description of Hawaiian invertebrates. Fresh-water fishes, amphibia and numerous invertebrates such as the giant snail *Achatina fulica* were introduced by man. MAYR (1943) and AMADON (1950) described the evolution of the bird fauna of Hawaii.

The geological preconditions on the **Galapagos** are like those on Hawaii. The fauna of the Galapagos is the best studied of island faunas (cf. JOHNSON & RAVEN 1973, MÜLLER 1973). It gave CHARLES DARWIN the first stimulus to his theory of evolution. It was here that he recognised the importance of spatial separation for the origin of species.

The vertebrate fauna of the Galapagos lacks amphibians or primary fresh-water fishes. The blind Brotulid fish *Caecogilbia galapagoensis* described by POLL and LELEUP (1965) lives in caves with varying pH (5.8–6.2) and strong oscillations in salinity (2.97–8.71‰) (cf. POLL & VAN MOL 1966, VAN MOL 1967).

The other species have very close connections with the Andean Pacific area and with Central America and, to a much lesser degree, with the Antilles.

The endemic iguana *Amblyrhynchus cristatus* occurs as closely interrelated subspecies on the islands of Narborough, Albemarle and Indefatigable within the 200 m depth contour. But on James the more strongly characterised subspecies *A. c. mertensi* occurs. As well as *Amblyrhynchus*, another endemic genus of iguana – *Conolophus* – lives on the Galapagos. The closest relatives of these genera are *Ctenosaurus*, *Cyclurus* and *Iguana* which are distributed in Central and South America. The remaining Galapagos reptiles are not generically

Island	Observed species		Area (km²)	Elevation (m)	Distance (km)		Area of adjacent island (km²)
	Total	Endemics			From nearest island	From Santa Cruz	
Baltra	58	23	25.09		0.6	0.6	1.84
Bartolomé	31	21	1.24	109	0.6	26.3	572.33
Caldwell	3	3	0.21	114	2.8	58.7	0.78
Champion	25	9	0.10	46	1.9	47.4	0.18
Coamaño	2	1	0.05		1.9	1.9	903.82
Daphne Major	18	11	0.34		8.0	8.0	1.84
Darwin	10	7	2.33	168	34.1	290.2	2.85
Eden	8	4	0.03		0.4	0.4	17.95
Enderby	2	2	0.18	112	2.6	50.2	0.10
Española	97	26	58.27	198	1.1	88.3	0.57
Fernandina	93	35	634.49	1494	4.3	95.3	4669.32
Gardner*	58	17	0.57	49	1.1	93.1	58.27
Gardner†	5	4	0.78	227	4.6	62.2	0.21
Genovesa	40	19	17.35	76	47.4	92.2	129.49
Isabela	347	89	4669.32	1707	0.7	28.1	634.49
Marchena	51	23	129.49	343	29.1	85.9	59.56
Onslow	2	2	0.01	25	3.3	45.9	0.10
Pinta	104	37	59.56	777	29.1	119.6	129.49
Pinzon	108	33	17.95	458	10.7	10.7	0.03
Las Plazas	12	9	0.23		0.5	0.6	25.09
Rabida	70	30	4.89	367	4.4	24.4	572.33
San Cristóbal	280	65	551.62	716	45.2	66.6	0.57
San Salvador	237	81	572.33	906	0.2	19.8	4.89
Santa Cruz	444	95	903.82	864	0.6	0.0	0.52
Santa Fé	62	28	24.08	259	16.5	16.5	0.52
Santa Maria	285	73	170.92	640	2.6	49.2	0.10
Seymour	44	16	1.84		0.6	9.6	25.09
Tortuga	16	8	1.24	186	6.8	50.9	17.95
Wolf	21	12	2.85	253	34.1	254.7	2.33

List of islands and plants species numbers (after JOHNSON & RAVEN 1973).

distinct from mainland forms and their closest relations are Andean Pacific faunal elements.

Such forms include the ground iguana *Tropidurus* and also the snake *Dromicus*. The latter can be divided into three groups or Formenkreise of *Dromicus biserialis*, *D. dorsalis* and *D. slevini*. The *Dromicus* species of the Galapagos can be derived from *Dromicus chamissonis* which has an Andean Pacific distribution, and, significantly, possesses scale pits like those characteristic of the Galapagos snakes. The distribution of the geckos of the Galapagos does not contradict that of other reptiles, though the lack of geckos on Narborough is striking. There is a decline in the number of species of geckos from Chatham in the east to Wenman in the west. Thus on Chatham the species *Phyllodactylus darwini*, *Ph. leei* and *Gonatodes collaris* occur while on all the other islands there is never more than a single species of *Phyllodactylus*. This westward decline in the number of species can be understood as indicating the direction of immigration of the Galapagos geckos.

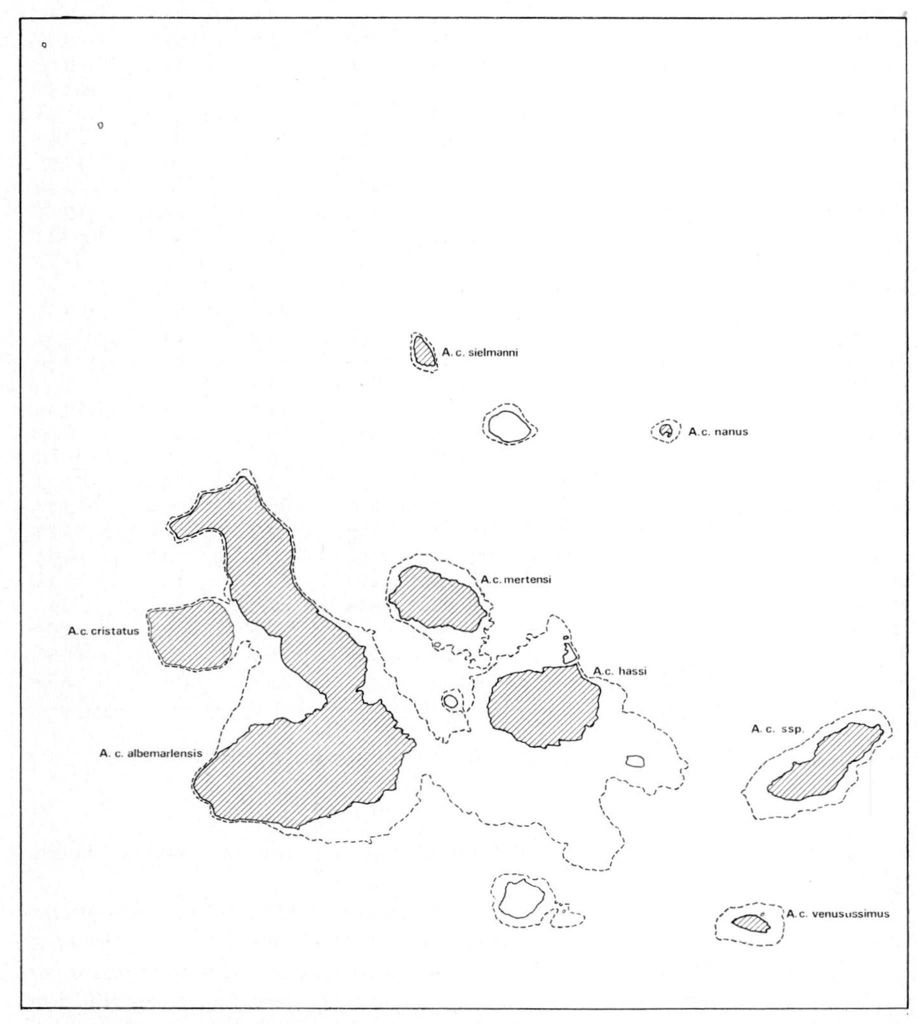

Fig. 97. The range of the Iguanid *Amblyrhynchus cristatus* on the Galapagos. The dashed line is the 200 m isobath. The islands on which *A. cristatus* occurs are cross-*hatched*.

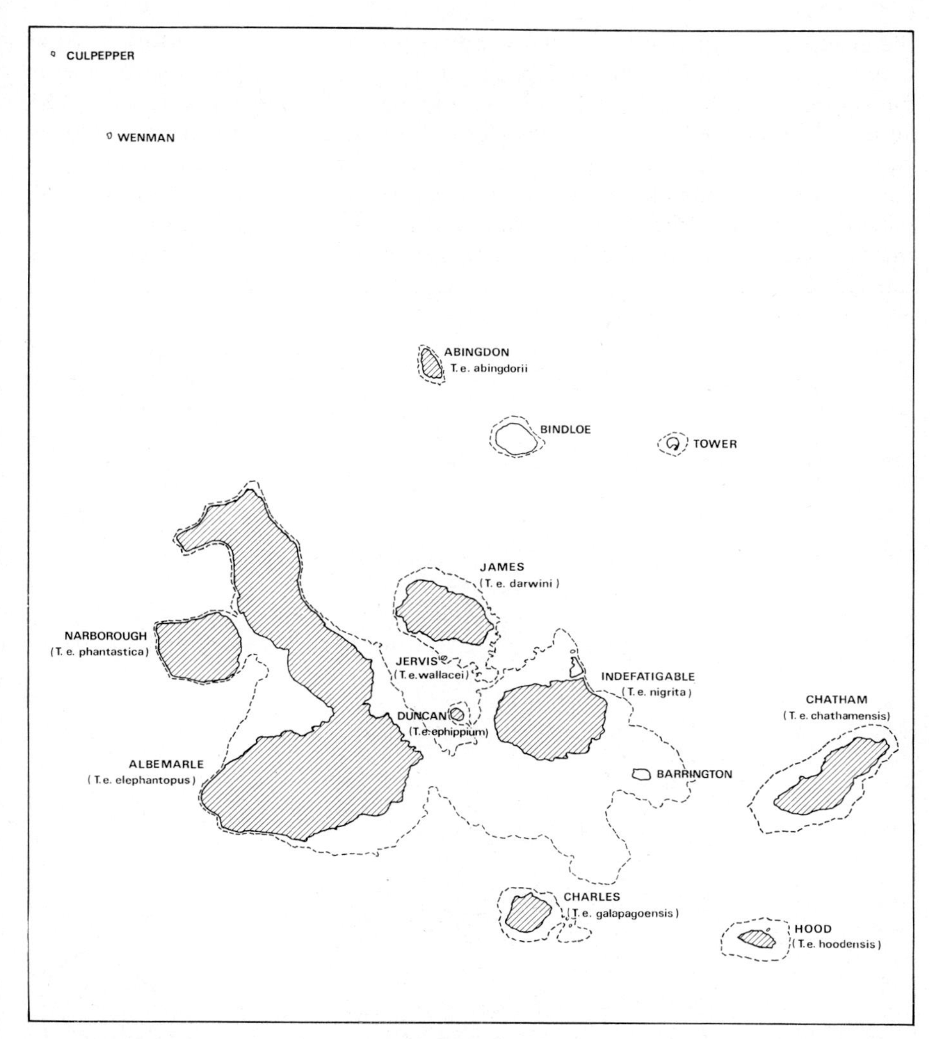

Fig. 98. Distribution of *Testudo* (*Geochelone*) *elephantopus* on the Galapagos (cross-hatched). Each island is inhabited by an endemic subspecies.

The giant tortoises of the Galapagos (*Geochelone elephantopus*) can also be derived from ancestors on the mainland of South America. It is worth mentioning in this connection that a giant tortoise – *Testudo cubensis* – is known from the Pleistocene of Cuba and is very like the Galapagos tortoises. *Testudo praestans* is known from the Pleistocene of Argentina. It has features of living species and also of *T. cubensis* and can be considered as possibly representing the group of origin. There are many indications that the giant tortoises of the Galapagos first reached the Galapagos at the beginning of the Pleistocene at the earliest, and that they arrived by swimming or floating across the sea. The 89 breeding birds of the Galapagos likewise immigrated across the sea. This is true for the flightless cormorant *Nannopterum harrisi*, the penguin *Spheniscus mendiculus*, the endemic seagulls *Creagrus furcatus* and *Larus fuliginosus*, the mocking thrushes of the genus *Nesomimus* and the Darwin finches which inhabit the Galapagos and Cocos islands as 14 species (LACK 1947). The Galapagos

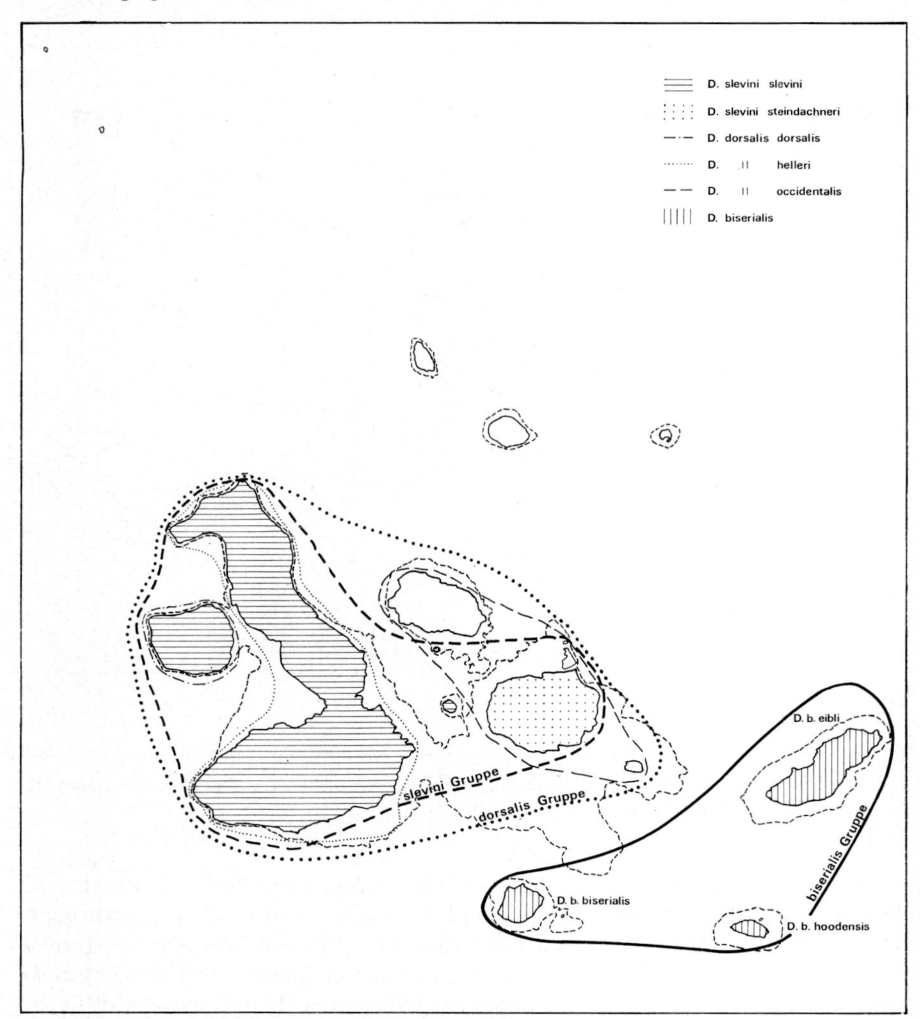

Fig. 99. Range of the snake genus *Dromicus* on the Galapagos.

finches probably share common ancestors with the continental genus *Tiaris*. A striking feature of the bird fauna of the Galapagos is the absence of humming birds.

The mammals of the Galapagos consist, apart from seals, of small forms only. They include: *Lasiurus brachyotis*, *L. cinereus*, *Rattus rattus*, *Mus musculus*, *Megalomys curioi*, *Nesoryzomys indefessus*, *N. darwini*, *N. swarthi*, *N. narboroughi*, *Oryzomys galapagoensis* and *O. bauri*.

Other small, geographically strictly isolated populations have a biogeographical importance similar to island biotas. Such populations include those of high mountains, which show an 'island distribution pattern' (cf. VARGA 1970, VUILLEUMIER 1970), and, above all, those of caves (cf. DE LATTIN 1941, 1967, CULVER 1973).

'In the origin of the **biota of caves** and subterranean waters the same causes operate as for the biota above ground. For it also the sole evolutionary factors are selection and isolation, acting on material provided by the natural heritable variation of organisms, which in turn arises solely by random mutation. It is not surprising that this variability in caves is different to above ground since certain degenerate types – eyeless, wingless or without pigment – are able to exist just as easily in caves as normal types. Above ground, on the other hand, they would almost always be wiped out by selection before they could reproduce. Because of its special peculiarities this out-of-the-way group of organisms is particularly suited to contribute, clearly and impressively, to our knowledge of the great evolutionary process by which all organisms, without exception, originate.' [1] (DE LATTIN 1941, p. 279).

Ecology, the developmental history of landscapes, and the evolution of animals can be made to illuminate each other mutually by the use of a method developed by REINIG (1939) and DE LATTIN (1957). This consists in working out and establishing the geographical positions of the **dispersal centres** of animals.

Dispersal centres are areas in which animals and plants survived unfavourable environmental conditions such as arid periods, glacial periods or the competitive pressure of other species. An area will only be important as a dispersal centre if the total life conditions acting in it do not, in the long term, cause the extinc-

1 'Auch für die Entstehung der Lebewelt der Höhlen und der unterirdischen Gewässer gelten keine anderen Ursachen als für diejenige oberirdischer Lebensräume. Auch für sie sind also die einzigen Entwicklungsfaktoren Auslese und Isolation, die an einem Material angreifen, das durch die natürliche erbliche Variation der Organismen geliefert wird, die ihrerseits ausschliesslich durch richtungsloses Mutieren zustande kommt. Dass diese Variabilität in Höhlen anders zusammengesetzt ist als oberirdisch, darf dabei nicht weiter wundernehmen, da hier bestimmte Degenerationsformen (mit Augen-, Pigment- und Flügelverlust) gleich gut zu existieren vermögen wie normale, während solche oberirdisch durch die Einwirkung der Auslese fast immer ausgemerzt werden, ehe sie zur Fortpflanzung kommen. So ist gerade diese abgelegene Gruppe von Lebewesen infolge ihrer besonderen Eigenart geeignet, einen besonders eindeutigen und eindrucksvollen Beitrag zu unserem Wissen von dem grossen Entwicklungsgeschehen, dem alle Lebewesen ohne Ausnahme ihre Entstehung verdanken, zu vermitteln'.

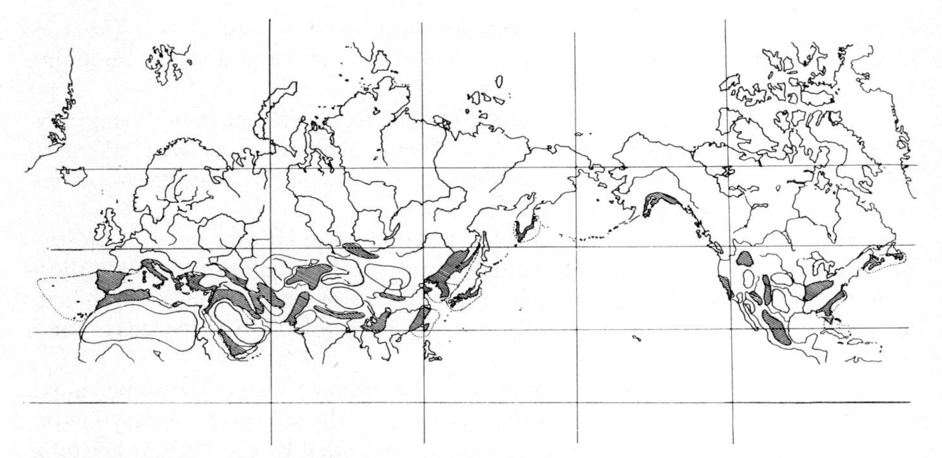

Fig. 100. Arboreal (= crosshatched) and eremial dispersal centres in the Holarctic realm (after DE LATTIN 1957).

tion of the life communities present in it. Thus dispersal centres are refuge areas by origin. For the duration of the unfavourable environmental conditions, populations exist in them as in areas isolated from other such areas and other populations. Thus one of the most important preconditions for the formation of species and subspecies is set in motion. This is the geographical separation of originally undivided populations.

The results got by working out and explaining dispersal centres are not only important for zoogeography. They contribute to an understanding of the evolution of organisms, throw light on the facts of climatic and earth history and thus lead to a deeper understanding of present-day landscape relationships. The importance of dispersal centres for the study of evolution is therefore just as great as their importance for geography. Thus a dispersal centre in the pampa of Argentina, inhabited by endemic species of animals adapted to an unforested landscape, argues against the view that the pampa was forest country before the arrival of man. But a precondition for the correctness of this assertion is the proof in the first place that the relevant groups of animals did disperse from this centre (MÜLLER 1973).

If the animals that occur are not represented by particular races this proof will be difficult to provide. We have already said that species can occur together in the same area without losing their specific characteristics. It is possible to imagine that a species arose in a place, died out in that place through a series of accidents but colonised one or more other areas in quite recent times. The present area of distribution of a species in no way needs to lie in its area of origin, nor even to make contact with it. It is therefore necessary to investigate an area of distribution further, to find out whether it lies near the area of origin

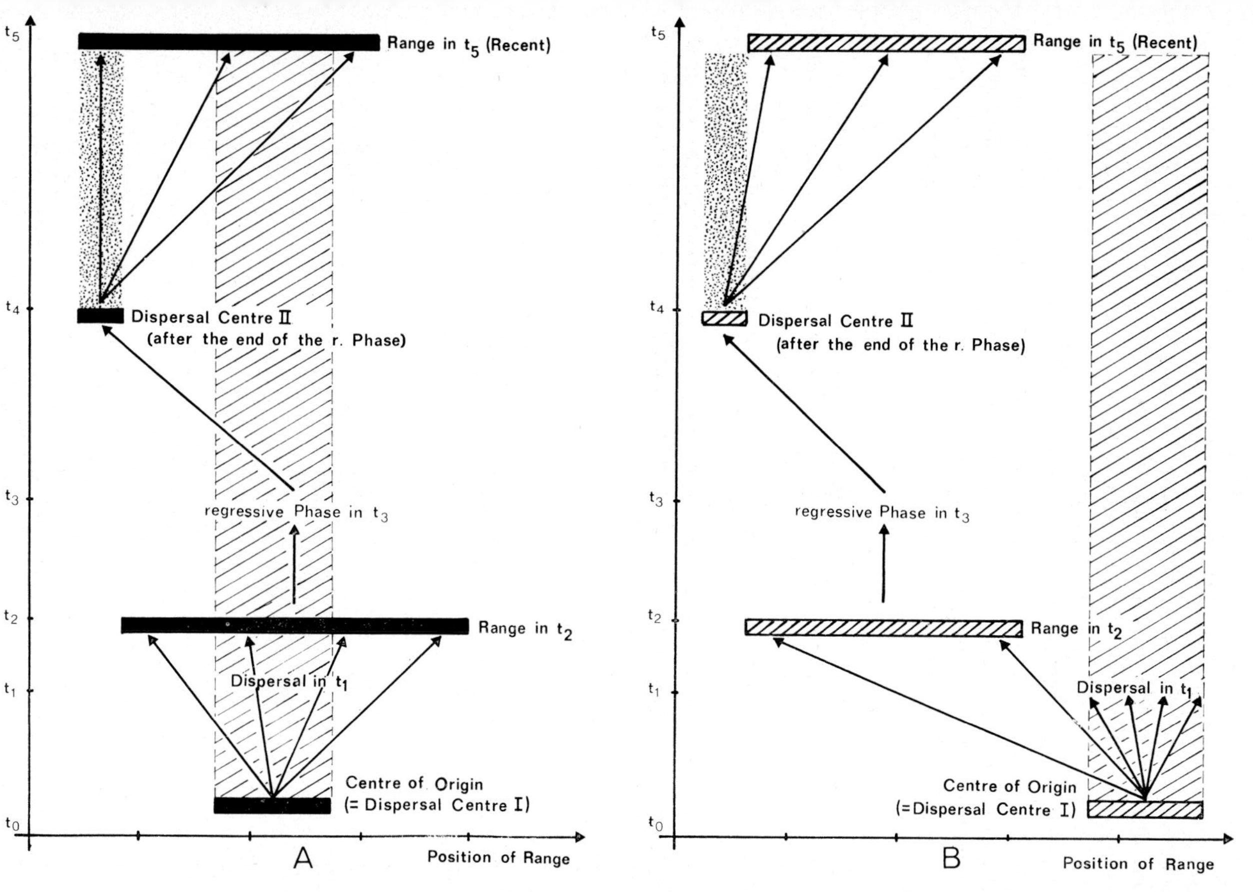

Fig. 101. Possible correlations between recent site of area, center of origin and dispersal centers which might occur in the course of the evolution of a taxon. The analysis of the dispersal centers can only provide information about the function of the most recent respective center of dispersal.

of a species (being plesiochorous), or whether, in the course of the evolution of a species it has become widely separated from the centre of origin (being apochorous). The solution to this question for higher systematic units such as genera or families depends on a complete description of their evolution, which can only be satisfactorily provided for a very few groups of animals. It therefore seems sensible to limit the study of dispersal centres in the first place to species, superspecies and geographical races or subspecies.

Subspecies and superspecies have one thing in common. Relative to the dispersal centre which last served as a refuge for a population they are mostly plesiochorous. The geographical range of a subspecies or a superspecies lies, with few exceptions, in the region of the dispersal centre of the population concerned.

By way of limitation it must be pointed out that formation of subspecies is not exclusively connected with geographical isolation of originally undivided populations. Subspecies can also arise by the migration of populations into regions with very different environmental conditions. This proposition was already elaborated by WAGNER (1868) in his **migration theory**.

It starts from the fact that differential evolution of populations will increase as soon as these migrate into new territories. Types present in the original populations that are better suited to the new conditions (i.e. pre-adapted) pene-

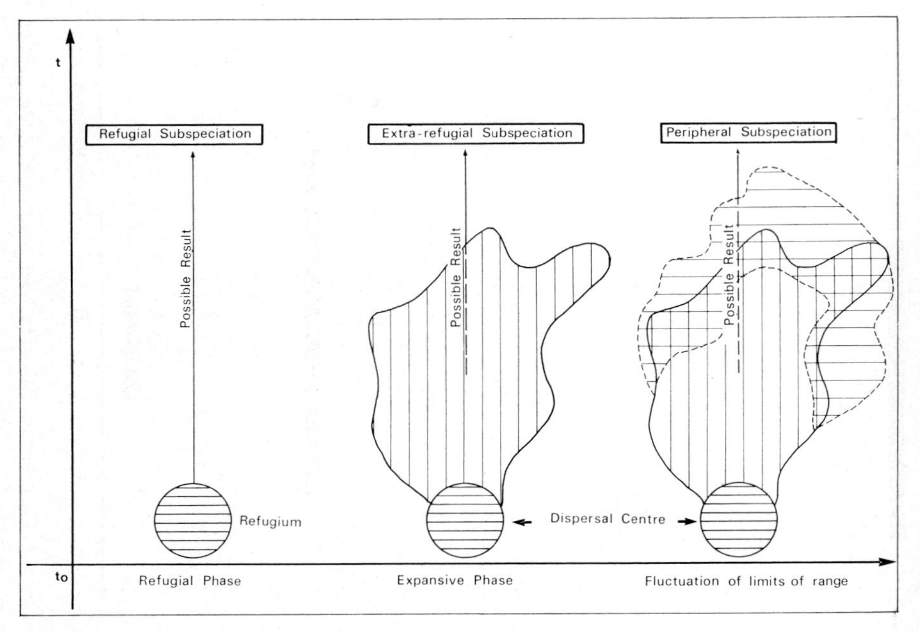

Fig. 102. Correlations between refugial phase, dynamics of the areas and Subspeciation.

trate the newly conquered territories quickly and soon affect the total picture of the population. This leads to very fast subspeciation. An example of this is provided by the house sparrow which was introduced towards the end of the last century into North America. The limitation just mentioned is a necessary precondition for a true estimate of the value of dispersal centres.

The working-out of dispersal centres requires several separate steps. The centres can be obtained by plotting the breeding areas of species, superspecies and subspecies on a map of a continent, a zoogeographical realm or of a smaller area. The various ranges do not often show common features at their boundaries but very often have in common an area of overlap, or nucleus of geographical range. These nuclei, however, are not necessarily centres of dispersal but merely centres of distribution (cf. COOK 1969, MÜLLER 1973). Whether they are also centres of dispersal, and thus represent the centres of preservation of faunas and floras during periods of unfavourable environmental conditions, can only be deduced by further study of the phylogenetic affinities of the faunas assigned to the centres. Every species possesses, or used to possess, a centre of dispersal that was also its centre of origin. In the course of evolution, however, the now-observed centre of dispersal can separate considerably from the centre of origin. The observed centres of dispersal merely represent areas in which populations

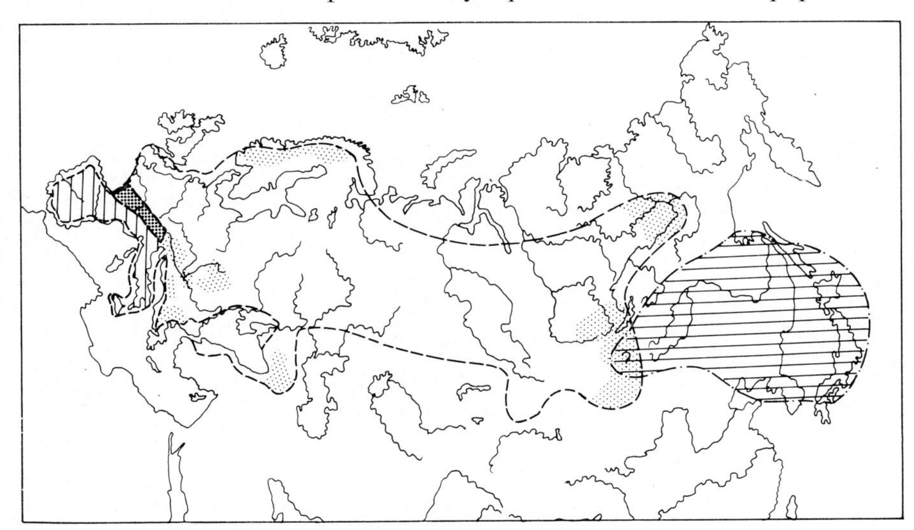

Fig. 103. Distribution and Hybridbelts from *Mellicta athalia* ROTT. Superspecies.

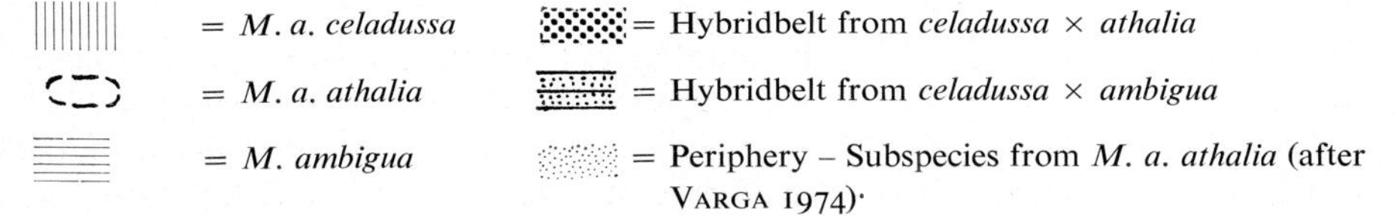

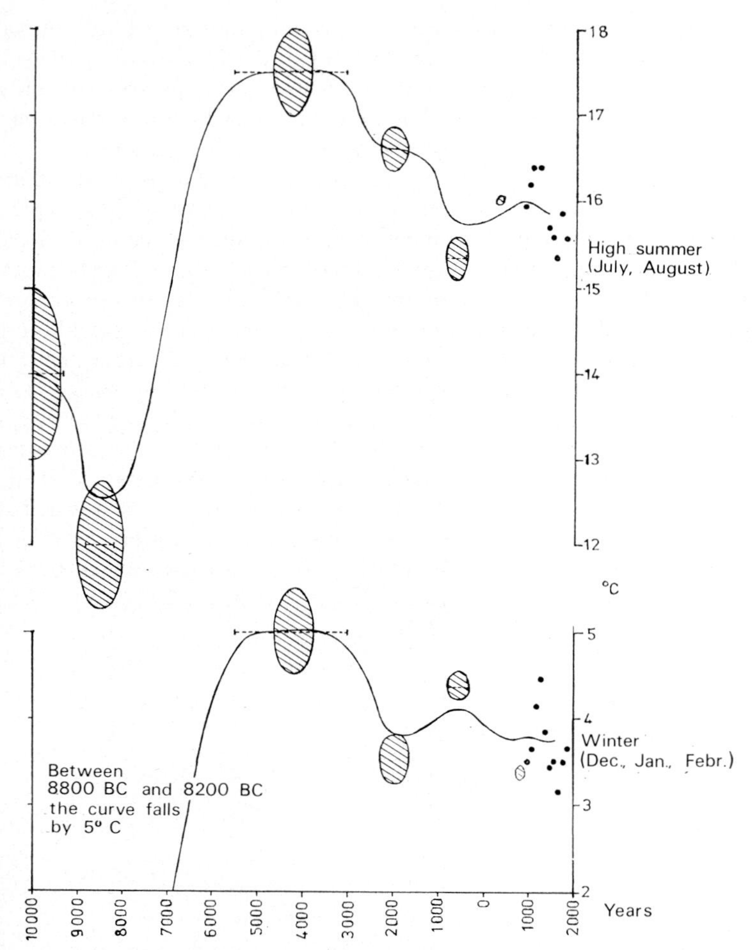

Fig. 104. Change of temperature during the last 12000 years (after LAMB, 1971); Interpolated line = connection between estimated 1000-year means;

⊢----⊣ = values for the last 10 centuries;

 = zone of uncertainty for the temperature values deduced from palaeobotanical data;

● = persistence of these temperatures according to C^{14}– dating (after LAMB, 1971).

survived the unfavourable environmental conditions that acted upon them last. In this respect a new type of dispersal centre is now forming all over the world. In these centres plants and animals are forced to survive the influence of man which has never before existed with its present intensity (BENNETT 1968, MÜLLER

1970, 1972, REED 1970, SUKOPP 1972, THOMAS 1956, VOS, MANVILLE & GELDER 1956).

The dispersal centres of phylogenetic and geographical importance which have been investigated up to now arose during the postglacial and glacial periods. In the Holarctic realm they stretch in a strip along the southern margin of the ice-free region of the Würm glaciation. These areas were studied by REINIG (1939) and DE LATTIN (1957). The animals and plants of Eurasia and North America found protection in them from the masses of ice that moved down from the North or out of the mountains. However it is not only a question of the climate becoming cooler in the last two million years. Precipitation relationships

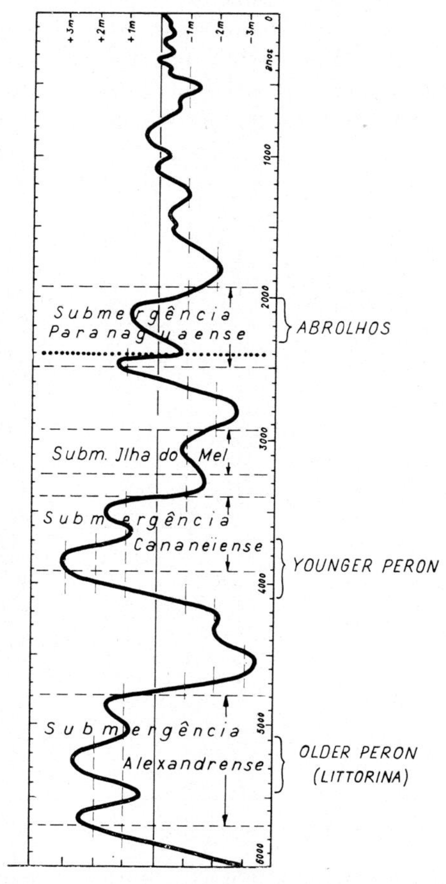

Fig. 105. The sequence of post-glacial variations of sea level on the Brazilian coast. The horizontal arrow indicates the end of the postglacial arid phase on the curitiba plateau (from MÜLLER 1973).

have also altered in many places. The effects of this within small areas is still not sufficiently known (cf. KROLOPP 1969). It is quite wrong to suppose that the tropics were climatically stable during the glacial periods (cf. BÜDEL 1970). It is now known that extensive arid and pluvial periods have alternated in the last two million years in South America, India, Africa, New Guinea and Australia.

By studies of the ranges of birds and mammals EISENTRAUT (1968, 1970) and MOREAU (1963, 1966, 1969) were able to prove extensive displacements of vegetation in tropical Africa. These were of critical importance for the formation of species in the African fauna.

Even in the early part of this century several workers demanded **vegetational fluctuations** in Africa to explain satisfactorily the speciation of savanna and rain-forest taxa (LÖNNBERG 1918, 1926, 1929; MOREAU 1931, CHAPIN 1932, BRAESTRUP 1935). Isolated islands of rain forest played an important rôle in these explanations. 'The faunas of these isolated patches show very close affinities to that of the great western hylaea, having many forms in common (e.g.

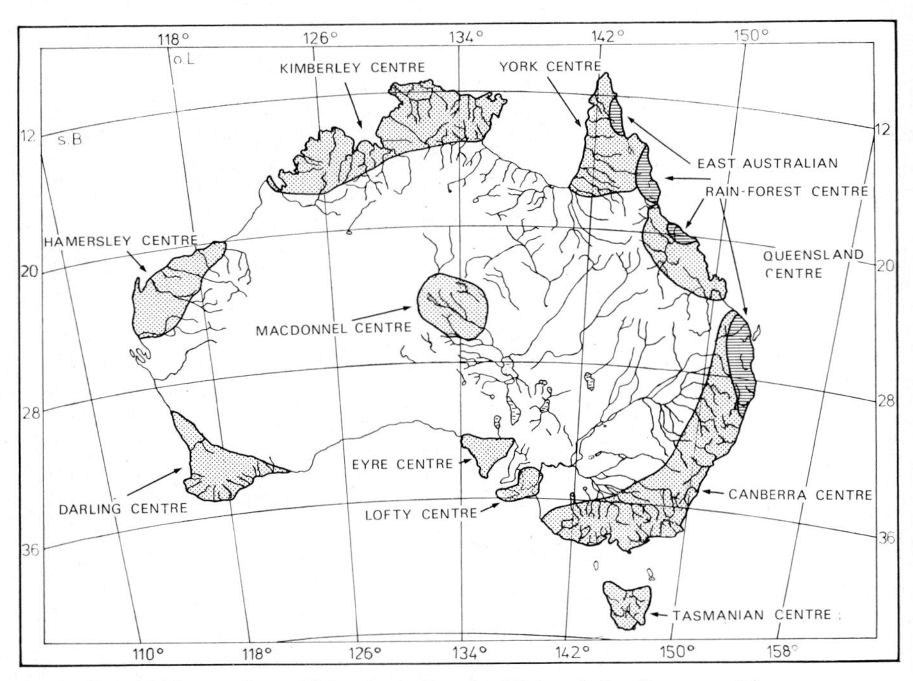

Fig. 106. Arid phase refuges of the Australian land fauna (after KEAST, 1961).

Anomalurus, Perodicticus) which impossibly could have crossed the vast intervening steppes. From this LÖNNBERG concluded, that the rain forest islands must have been once continuous with the main western forest owing to a moister climate'. (BRAESTRUP 1935).

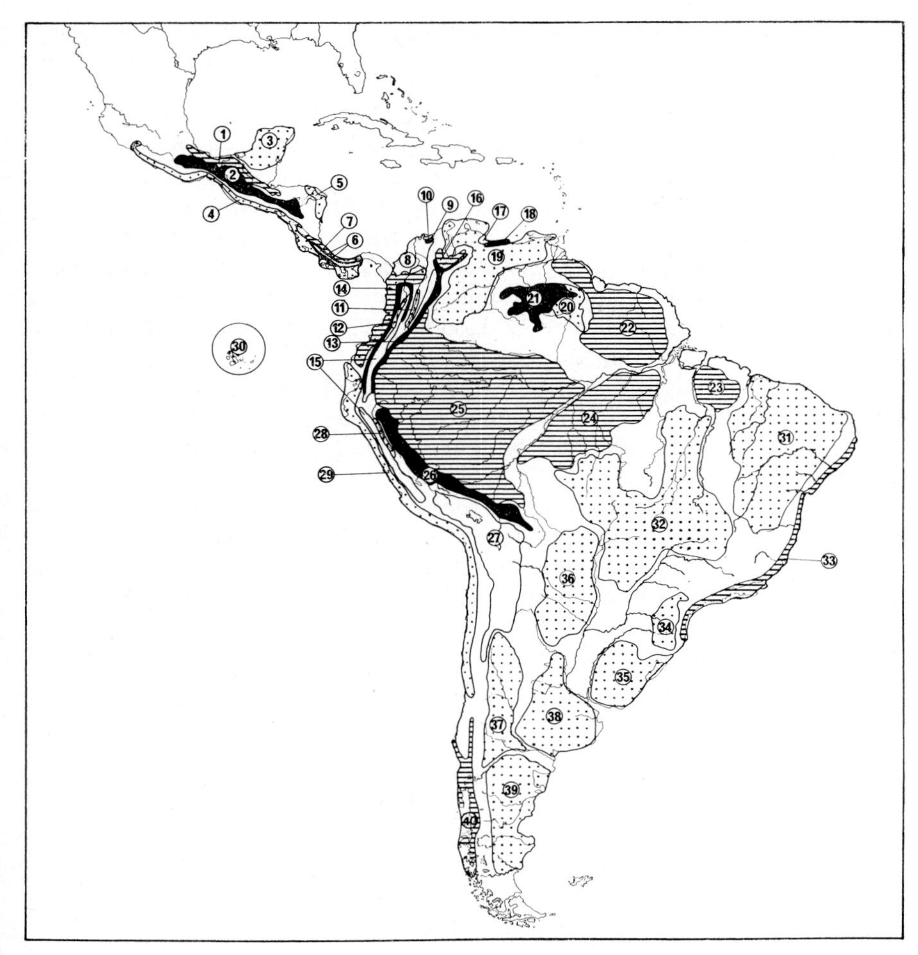

Fig. 107. Dispersal centres of terrestrial vertebrates in the Neotropical realm. Black = montane forest centres; White = oreal centres; obliquely hatched = rain forest centres; stippled = non-forest centres.

GENTILLI (1949) and KEAST (1959, 1961, 1968) have also supposed that much of the speciation and subspeciation of Australian animals was connected with the isolation of populations round the edges of the continent in dry climatic conditions. PIANKA (1972) demanded 'habitats fluctuating in space and time' for the evolution of Australian reptiles (for plants cf. CROCKER 1959).

The age of the arid period which caused these Australian dispersal centres is still not certainly established. Pollen-analytical and geological studies confirm the biogeographical results in both Africa and Australia. Corresponding to the arid periods and displacements of vegetation in Africa and Australia there must have been similar events in South America. It is nevertheless striking that campo expansions occurred in South America even in the Post glacial and, in the course of these, campo cerrado animals invaded Amazonia. The faunas of South American islands can be used as time indicators for these territorial changes in central and eastern South America.

This expansion of the campo can be proved to have lasted from 6000 to 2400 B.C. and led to an expansion of the restinga from Cabo Frio as far as the Rio Grande do Sul and to an arid period in north east Brazil. The thoroughfares used by species adapted to open landscapes were broken up by a renewed expansion of forest that set in about 2400 B.C. This expansion of forest, with a few exceptions, has continued to the present.

The campo islands within the Amazonian rain forest have a characteristic campo fauna and flora and were connected together by a 'climatic-bridge' or corridor of savanna. They must be seen as relicts of this recent arid period (MÜLLER 1973).

The position of the South American dispersal centres, especially of the forest centres, was critically influenced by the arid periods. It is remarkable to observe that regions of hybrid belts of subspecifically distinct forest populations lie in the areas of these former thoroughfares of campo which have only been won

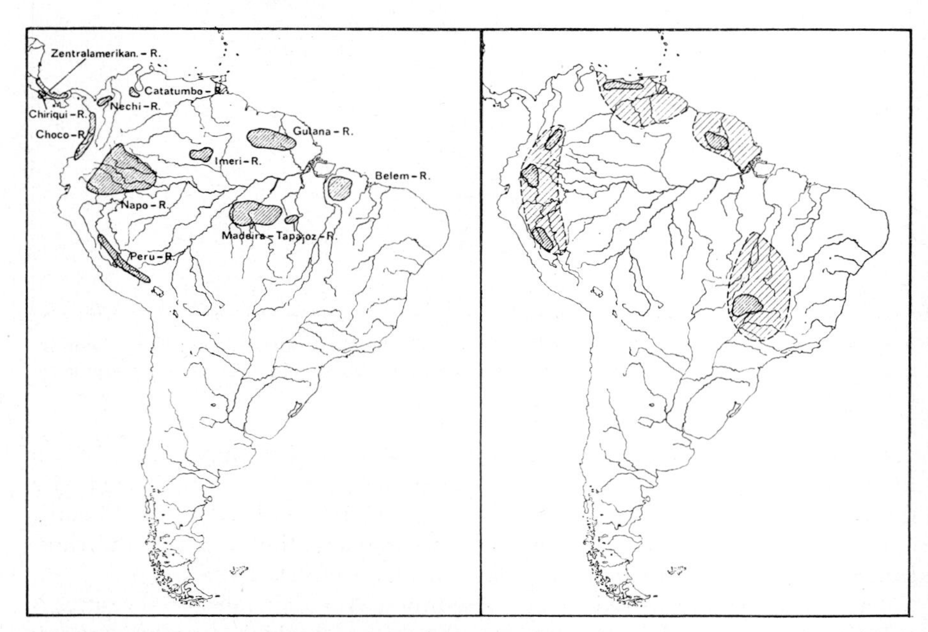

Fig. 108. Positions of refuges as worked out by HAFFER (1969) for Amazonian rain-forest birds (left) and by VANZOLINI & WILLIAMS (1970) for the *Anolis chrysolepis* group (right).

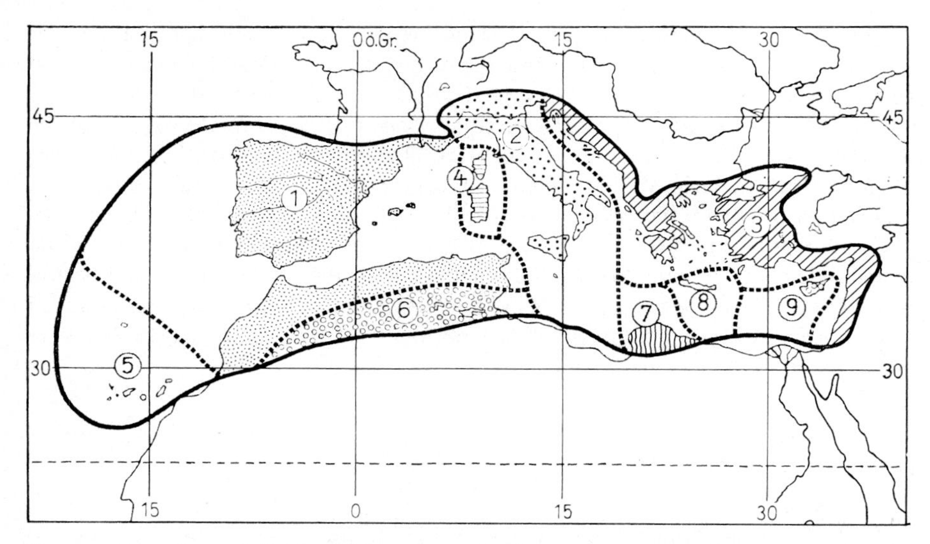

Fig. 109. Secondary subdivision of the primary Mediterranean centre, after DE LATTIN (1967); Subcentres as follows; 1. Atlantomediterranean, 2. Adriatomediterranean, 3. Pontomediterranean, 4. Tyrrhenian, 5. Canaries, 6. Mauretanian, 7. Cyrenaican, 8. Cretan, 9. Cyprus.

back by the forest within the last 4000 years. The beginning and end of the postglacial arid period were marked by wetter conditions with expansions of rain forest. These expansions led to the isolation of populations of animals adapted to open landscapes. The campo cerrado, from the beginning of the postglacial arid phase up to the present day, has hindered or totally prevented any gene exchange between the rain forest population of the Brazilian coastal forest and Amazonia. But about 7000 BC it must in particular places have been considerably more forested than now.

The Andes, which only reached their present altitude in the later Tertiary and the Pleistocene, were glaciated in the Würm Glaciation in several areas, like the highest mountains of New Guinea and Africa. With the strong increase in temperature during the postglacial period there occurred an increased subdivision of the oreal and montane forest fauna as a result of displacements in altitude. These fauna had preferred to live in lowland environments up until about 11000 BC.

The large number of Neotropical dispersal centres can be connected with the unparalleled richness in species of South American biomes. Zoogeographers have been able to show that a large part of the species-diversity of the Neotropical rain forests can be ascribed to varying isolation of the forest area during arid climatic periods.

The recent climatic and landscape history of South America, deduced from the Neotropical dispersal centres, only represents the start of the study of South

America from the viewpoint of historical zoogeography. The dispersal centres also have their own history with times of origin and periods of dissolution. Also the dispersal centres are only an outer framework and can be subdivided.

The Mediterranean dispersal centre studied by DE LATTIN (1957) can be divided into at least nine smaller centres. Taking a single one of these, for example the Adriatic-Mediterranean subcentre that includes Italy and Sicily, and studying the refuges of warmth- and cold-loving species, a still finer subdivision becomes necessary. It then appears that a warmth-loving reptile fauna was divided by the glaciation of the Apennines during the Würm glacial period into small refuges near the coast. These former refuge areas can be deduced from the subspecific evolution of the populations of Italian lizards (*Lacerta sicula*, *L. muralis*, *L. viridis*, etc.).

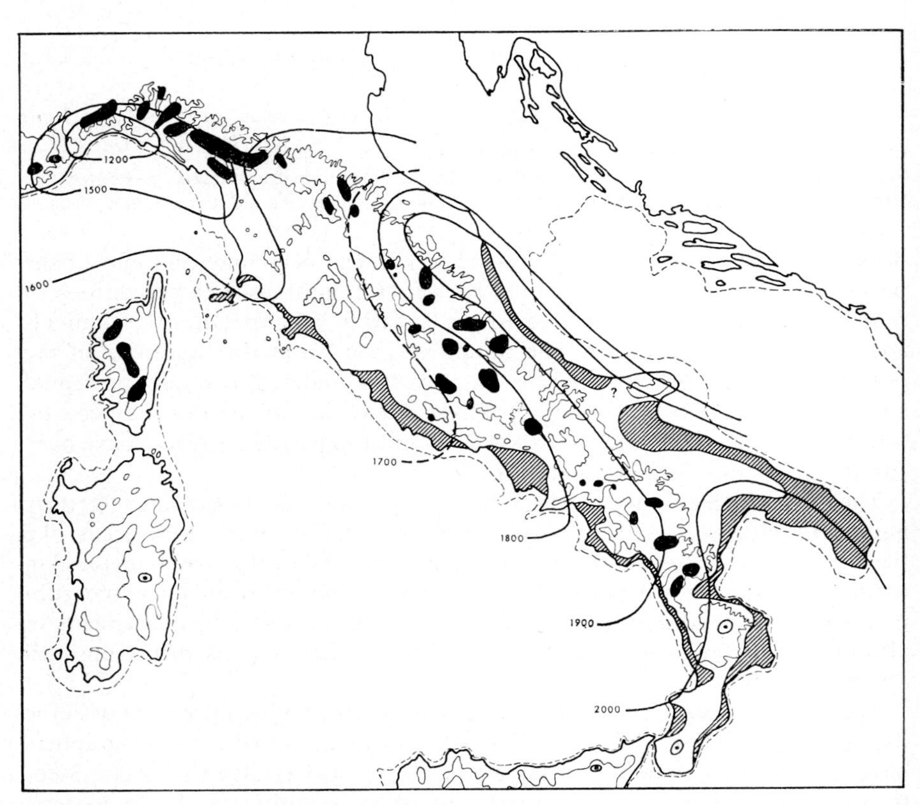

Fig. 110. The Adriatic – Mediterranean dispersal centre during the Würm glacial (after SCHNEIDER 1971). The Apennines were glaciated as far as so thern Italy (black). The snow lines moved downhill (after MESSERLI 1967). Warmth-loving elements lived in small refuges near the coast (from MÜLLER 1972).

I have already said that dispersal centres are the starting point for historical zoogeography. The totality of a life-community ascribed to a dispersal centre is made up of many layers with groups of very different ages. These different layers can only be understood if much older processes of evolution and range-history are considered.

This leads to the developmental history of whole continents and even of the whole earth. THENIUS (1972) believes that: 'the present distribution of animals and plants cannot be properly explained either by the present-day distribution of water and land, or by the ecological factors now active.'[1]

But to try to advance zoogeography by applying the results of geophysics, oceanography, geology and palaeontology only seems justified to me if zoogeography, without prejudice and using its own methods, can present causal connections rather than coincidences.

Animals and plants will only throw light on the development of the earth, if we know their own evolution and their ecological valency. Resemblances signify nothing if they do not indicate phylogenetic relationships. The path leading, through Gondwanaland or over the South Pacific land connections in the region of Archinotis, can only be correctly followed if, as well as working out ecology and recent changes of range of species, we can also set up a phylogenetic systematics by means of which the evolutionary history of species is transformed into a history of the evolution of the world. So long as those basic facts are not given enough attention, so long will zoogeography be given over to pure coincidence. Science only begins when coincidence can be distinguished from causality.

1 'die gegenwärtige Verbreitung der Tiere und Pflanzen weder durch die heutige Verteilung von Wasser und Land noch durch die derzeit wirksamen ökologischen Faktoren hinreichend zu erklären ist'.

REFERENCES

ADAMS, R. (1970): Contour Mapping and Differential Systematics of Geographic Variations. Syst. Zool.

ALLEN, J. A. (1878): The geographic distribution of mammals. *Bull. U.S. Geol. Geo. Survey* 4: 339–343.

AMADON, D. (1950): The Hawaiian honeycreepers (Aves, Drepaniidae). *Bull. Amer. Mus. Nat. Hist.* 95: 151–262.

ANDER, K. (1942): Die Insektenfauna des baltischen Bernsteins nebst damit verknüpften zoogeographischen Problemen. *Kung. Fysiogr. Sällsk. Handl. N.S.* 53, No 4 pp. 83.

ANDRIASHEV, A. P. (1965): A general review of the antarctic fish fauna. In: Biogeography and Ecology in Antarctica, 491–550, The Hague.

ANT, H. (1969): Die Malakologische Gliederung einiger Buchenwaldtypen in Nordwestdeutschland. *Vegetatio* 18: 374–386.

ARLDT, T. (1907): Die Entwicklung der Kontinente und ihrer Lebewelt. Leipzig.

ASPÖCK, H. (1965): Studies of Culicidae (Diptera) and consideration of their role as potential vectors of Arboviruses in Austria. XII Int. Cong. Entomol. London 767–769.

AUDY, J. R. (1956): The role of Mite Vectors in the Natural History of Scrub Typhus. 10. *Int. Congr. Entomol. Montreal* 3: 639–649.

AXELROD, D. J. (1958): Evolution of the Madro-Tertiary. *Geoflora. Botan. Rev.* 24: 433–509.

AXELROD, D. J. (1972): Plate tectonics and problems of Angiosperm History. XVII Congrès internat. Zool. Biogéographie et liaisons inter-continentales au cours du mésozoique. Monte Carlo (Manuskript).

AXELROD, D. J. & TING, W. S. (1961): Early Pleistocene floras from the Chagopa surface, southern Sierra Nevada. *Univ. Calif. Publ. Geol. Sci.* 39: 119–194.

BACH, W. (1972): Urban air pollution: climatological modelling. International Geography 1972; 22. *Int. Geogr. Congr.* 1: 129–131.

BAKER, R. H. (1960): Mammals of the Guadina Lava Field, Durango, Mexico. *Mich. State Univ. Biol. Ser.* 1: 305–327.

BALACHOWSKY, A. S. (1972): Entomologie appliquée à l'Agriculture. 2 Lépidoptères. Masson u. Cie., Paris.

BALOGH, J. (1958): Lebensgemeinschaften der Landtiere. Vg. Ungar. Akad. Wiss., Budapest pp. 560.

BANARESCU, P. (1967): Die zoogeographische Stellung der Fauna der unteren Donau. *Hidrobiologica* 8: 151–162. Bucarest.

— (1970): Principi si probleme de zoogeografie. Acad. Rep. Soc. Rom., Bucarest.

BANARESCU, P. & BOSCAIN, N. (1973): Biogeografie. Edit. Stintifica, Bucarest.

BANTA, J. E. & FONAROFF, L. S. (1969): Some considerations in 'The Study of geographic distribution of Diseases'. *Professional Geogr.* 21: 87–92.

BARNETT, H. C. (1960): The incrimination of Arthropods as vectors of disease. 11. Int.Kongr. Entomol. Wien, 341–345.

BARTENEW, A. (1932): Über einige Grundfragen der Zoogeographie. *Zool. Zh.* 2: 23–38.

BARTHOLOMEW, J. G., W. E. CLARK, & P. H. GRIMSHAW (1911): Atlas of zoogeography. Bartholomew's, London.

BAUER, K. (1961): Studien über Nebenwirkungen von Pflanzenschutzmitteln auf Fische. *Mitt. Biol. Bundesamt. Land u. Forstwirtsch.* 105.

BEAUFORT, L. F. DE (1951): Zoogeography of the land and inland waters. Sidgwick and Jackson London. pp.208.
BECKER, F. (1972): Die Bedeutung der Orographie in der medizinischen Klimatologie. Geogr. Taschenbuch 342–356.
BECKER, J. (1972): Art und Ursachen der Habitatbindung von Bodenarthropoden (Carabidae, Diplopoda, Coleoptera, Isopoda) xerothermer Standorte in der Eifel. Dissertation, Köln.
BEDDARD, F. E. (1895): A text-book of Zoogeography. Cambridge.
BENNET, CH. F. (1968): Human influences on the Zoogeography of Panama. Ibero-Americana 51: 1–112, Univ. Calif. Press, Berkeley and Los Angeles.
BENSON, S. B. (1933): Concealing coloration among desert rodents of southwestern United States. *Univ. Calif. Publ. Zool.* 40: 1–70.
BENTON, A. H. (1959): Observations on host perception in fleas. *J. Parasit. Lancaster, Pa.* 45: 614.
BERG, L.S. (1933): Die bipolare Verbreitung der Organismen und die Eiszeit. *Zoogeographica* 1 (4): 449–484.
BERNDT, R. & DANCKER, P. (1966): Die Expansion der Türkentaube (Streptopelia decaocto) – eine notwendige Folge ihrer Populationsdynamik. *Vogelwelt* 87 (2): 48–52.
BERNDT, R. & KLUYVER, N. (1967): Die Kohlmeise, Parus major, als Invasionsvogel. Vogelwarte.
— (1968): Terms, Studies and Experiments on the Problems of Bird Dispersion. *Ibis* 110: 256–269.
BERNDT, R. & STERNBERG, H. (1969): Alters- u. Geschlechtsunterschiede in der Dispersion des Trauerschnäppers (Ficedula hypoleuca). *J. Ornithol.* 110 (1): 22-26.
— (1969): Über Begriffe, Ursachen und Auswirkungen der Dispersion bei Vögeln. *Vogelwelt* 90 (2): 41–53.
BESUCHET, C. (1968): Répartition des insectes en Suisse. Influence des glaciations. *Mitt. schweiz. ent. Ges.* 41: 337–340.
BIGARELLA, J. J. et al. (1969): Processes and Environments of the Brazilian Quaternary. In: The Periglacial Environment. Montreal.
BIRIUKOV, V. J. (1944): Acclimatization of Gambusia in the Kharkov district. *Med. Parasitol.* 13.
BLAIR, W. F. (1943): Ecological distribution of mammals in the Tularosa Basin, New Mexico. *Contrib. Lab. Vert. Biol., Univ. Mich.* 20: 1–24.
BLANC, CH. P. (1971): Les reptiles de Madagascar et des îles voisines. *Annales Univ. Madagascar* 8: 95–178.
BLANFORD, W. T. (1890): Anniversary address to the Geological Society. *Proc. Geol. Soc. London* 1890: 43–110.
BOBRINSKIJ, N. A. (1927): Zoogeografie i ekologie. Moscow.
BODENHEIMER, F. S. (1959): The Uniqueness of Australia in Biology. In: Biogeography and Ecology in Australia. Verl. Junk, Den Haag.
BOGDANOVICH, I. N. (1935): The reproduction and acclimatization of Gambusia in Turkmenistan. *Med. Parasitol.* 4.
BRAESTRUP, F. W. (1935): Remarks on climatic change and Faunal Evolution in Africa. *Zoogeographica* 2 (4): 484–494.
BRAUER, A. (1914): Tiergeographie. In: Die Kultur der Gegenwart, Leipzig.
BROCK, T. D. (1967): Life at high temperatures. *Science* 158: 1012–1019.
— (1969): Microbial growth under extreme environments. *Symp. Soc. gen. Microbiol.* 19: 15–41.
BROOKS, R. R. (1972): Geobotany and Biogeochemistry in Mineral Exploration. Harper u. Row Publ., New York, San Francisco, London.
BROWN, C. E. (1964): A machine method for mapping insect survey records. Contr. 1103. Forest Ent. & Pathol. Branch, Dept. For. Ottawa.

BRUES, CH. T. (1928): Studies on the fauna of hot springs in the western United States and the biology of thermophilous animals. *Proc. Amer. Acad. Arts Sci.* 63: 129–228.
BRUIJNING, C. F. A. (1956): The occurrence of some Insect-Borne Diseases in Suriname in relation to the distribution of their vectors. 10. Int. Congr. Entomol. Montreal 3: 655.
BRUNDIN, L. (1965): On the real nature of Transatlantic relationships. *Evolution* 19: 496–505.
— (1966): Transantarctic relationships and their significance, as evidenced by chironomid midges, with a monograph of the subfamilies Podonominae and Aphroteniinae and the austral Heptagyiae. Kungl. Svenska Vetenskapsakademeins Handlinger. Fj. Ser. 11.
— (1972): Circum-Antarctic distribution patterns and continental drift. XVII Congrès internat. Zool. Biogéographie et liaisons inter-continentales au cours du Mésozoique. Monte Carlo (Manuskript).
BÜDEL, J. (1970): Die Eiszeit und wir. *Journ. Muße und Genesung* 3 (1).
BUHL, A. (1969): Punktkartierung und Rasterkartierung im Bereich des Kartierungsgebietes der Arbeitsgemeinschaft Mitteldeutscher Floristen. – Ein pflanzengeographisch-kartographischer Vergleich. *Wiss. Z. Univ. Halle-Wittenberg* 18: 475–480.
BYERS, G. W. (1969): Evolution of Wing Reduction in Crane Flies (Diptera: Tipulidae). *Evolution* 23 (2): 346–354.
CADBURY, D. A., HAWKENS, J. G. & READETT, R. C. (1971): A Computer-Mapped Flora. A Study of the County of Werwickshire. Acad. Press, London und New York.
CANDOLLE, A. P. DE (1855): Géographie botanique. Masson, Paris and Geneva. pp. xxxii + 1366.
CAPPETTA, H., RUSSELL, D. E. & BRAILLON, J. (1972): Sur la découverte de Characidae (Pisces, Cypriniformes) dans l'éocène inférieur français. XVII Congrès internat. Zool. Biogéographie et liaisons inter-continentales au cours du Mésozoique. Monte Carlo (Manuskript).
CARLQUIST, S. (1965): Island life. Doubleday, New York. 451 pp.
— (1966): The biota of long-distance dispersal I. Principles of dispersal and evolution. *Quart. Rev. Biol.* 41: 247–270.
— (1966): The biota of long-distance dispersal. II. Loss of dispersibility in Pacific Compositae. *Evolution* 20: 30–48.
CAROL, H. (1963): Zur Theorie der Geographie, Mitt. Österr. Geogr. Ges.
CARPENTER, J. R. (1939): The Biome. *Amer. Midl. Natural.* 21 (1).
CASPERS, H. & MANN, H. (1961): Bodenfauna und Fischbestand in der Hamburger Alster. Abhdl. u. Verhandl. Naturwiss. Ver. Hamburg 5: 89–110.
CHACKO, P. J. (1948): On the habits of the exotic mosquito-fish Gambusia affinis (Baird & Girard) in Waters of Madras. *Curr. Sci.* 17.
CHALONER, W. G. & MEYEN, S. V. (1973): Carboniferous and Permian Floras of the Northern Continents. In: Atlas of Palaeobiogeography. Elsevier Scient. Publ. Comp. Amsterdam, London, New York.
CHAMBERLAIN, R. W. (1956): Virus-Vector-Host Relationships of the American Arthropod-Borne Encephalitides. 10. Inter. Congr. Entomol. Montreal 3: 567–572.
CHAPIN, J. P. (1932): The birds of the Belgian Congo. *Bull. Amer. Mus. Nat. Hist.* 65: 1–756.
CHIKISHEV, A. G. (1973): Landscape indicators. New Techniques in Geology and Geography. Consult Bureau, New York, London.
CLARK, A. H. (1949): The Invasion of New Zealand by people, plants and animals. New Brunswick.
CLEMENTS, F. E. & V. E. SHELFORD (1939): Bio-ecology. Wiley, New York, pp. 825.
CODY, H. L. (1973): Coexistence, Coevolution and convergent evolution in Seabird communities. *Ecology* 54 (1): 31–44.
COLBERT, E. H. (1972): Early Triassic Tetrapods and Gondwanaland. XVII. Congrès internat. Zool. Biogéographie et liaisons inter-continentales au cours du Mésozoique. Monte Carlo (Manuskript).

COLINVAUX, P. (1973): Introduction to Ecology. John Wiley & Sons, Inc. New York, London, Sydney, Toronto.
COLLIER, B. D., COX, G. W., JOHNSON, A. W., MILLER, PH. C. (1973): Dynamic Ecology. Prentice-Hall, Inc., Englewood Cliffs, N.J.
COOK, R. E. (1969): Variation in species density of North American birds. *Syst. Zoology.*
CRACRAFT, J. (1972): Mesozoic dispersal of terrestrial Faunas around the Southern End of the world. XVII. Congrès internat. Zool. Biogéographie et liaisons inter-continentales au cours du Mésozoique. Monte Carlo (Manuskript).
CRALLEY, L. V. (1972): Industrial Environmental Health. Acad. Press. New York and London.
CROCKER, R. L. (1959): Past climatic fluctuations and their Influence upon australian Vegetation. In: Biogeography and Ecology in Australia. Verl. Junk, Den Haag.
CROMBIE, A. C. (1946): Further experiments on Insect Competition. *Proc. Roy. Soc. London* 133: 76–109.
CROWE, CH. R. (1966): Application of Automation in Rhopalocera research. *J. Lepid. Soc.* 20 (1): 1–12.
CRUSAFONT-PAIRO, M. AND PETTER, F. (1964): Un Murine geant des Iles Canaries Canariomys bravoi Gen. nov., sp. nov. (Rongeurs, muridés). *Mammalia* 28: 607–612.
CULLEY, M. (1971): The Pilchard; Biology and exploitation. Pergamon Press, Oxford, New York, Toronto, Sydney, Braunschweig.
CULVER, D. C. (1973): Competition in spatially heterogeneous systems an analysis of simple cave communities. *Ecology* 54 (1): 102–110.
CUMBERLAND, K. B. (1940): Primitive Vegetation of New Zealand. *Geographical Rev.* 30 (4), New York.
— (1941): A Century's change: Natural to Cultural Vegetation in New Zealand.
— (1962): 'Climatic change' or Cultural Interference? New Zealand in Moahunter Times. Land and Livelihood. *N.Z.G. Soc., Christchurch* 88–142.
DAHL, F. (1921): Grundlagen einer ökologischen Tiergeographie. Jena.
— (1925): Tiergeographie. In: KENDE, Encycl. der Erdkunde, Leipzig, Wien.
DALENIUS, P. (1965): The Acarology of the Antarctic Regions. In: Biogeography and Ecology in Antarctica, 414–430, Junk, The Hague.
DAMMERMAN, K. W. (1922): The fauna of Krakatau, Verlaten island and Sebesy. *Treubia* 3: 61–111.
— (1948): The fauna of Krakatau, Amsterdam.
DANSEREAU, P. (1957): Biogeography, an ecological perspective. Ronald, New York, pp. 394.
— (1968): Les structures de végétation. Finisterra. *Rev. Port. Geogr.* 3: 147–174.
— (1970): Megalopolis: resources and prospect. Challenge for Survival 1–33; Columbia Univers. Press, New York und London.
— (1971): Dimensions of environmental quality. *Sarracenia* 14: 1–109. Montréal.
DARLINGTON, P. J. (1957): Zoogeography, an ecological perspective. New York.
— (1970): A practical Criticism of Hennig-Brundin 'Phylogenetic Systematics' and Antarctic biogeography. *System. Zool.* 19: 1–18.
DARWIN, CH. (1832): The structure and distribution of Coral reefs. London.
— (1876): Struktur und Verteilung der Korallenriffe. Stuttgart.
DAVIES, J. B. (1965): Studies on the Dispersal of Leptoconops bequaerti KIEFER (Diptera: Ceratopogonidae) by means of wind traps. XII Int. Congr. Entomol. London. 754–755.
DEEVEY, E. S. (1949): Biogeography of the Pleistocene I. Europe and North America. *Bull. Geol. Soc. Am.* 60: 1315–1416.
DEN BOER, P. J. (1973): Das Überleben von Populationen und Arten und die Bedeutung von Umweltheterogenität. Verhdl. Dtsch. Zool. Ges. Mainz 66: 125–136.
DEVER, A. G. E. (1972): Training a medical geographer. Int. Geography Montreal 1972: 1212–1214.
DIELS, L. (1896): Vegetationsbiologie von Neuseeland. *Englers bot. Jb.* 22. Leipzig.

DIERL, W. (1970): Grundzüge einer ökologischen Tiergeographie der Schwärmer Ostnepals (Lepidoptera-Sphingidae). *Khumbu Himal* 3: 313–360.
DILCHER, D. L. (1969): Podocarpus from the Eocene of North America. *Science* 164: 299–301.
DÖRRE, A. (1926): Kälteindustrie und Mäuseschäden. *Mitt. Ges. Vorratsschutz* 2: 23–24.
DORST, J. (1962): The migration of birds. Riverside Press, Cambridge, Mass. pp. xiv+476 (translation of Les migrations des oiseaux, 1956, Payot, Paris).
DUVIGNEAUD, P. et al. (1962): L'écosystème. L'écologie, science moderne de synthèse. Brüssel.
EDWARDS, R. W. & GARROD, D. J. (1972): Conservation and Productivity of Natural Waters. Acad. Press, New York.
EHRIG, F. R. (1973): Zum Problem der Macchien am Beispiel Korsikas. Mitt. Geogr. Ges. München 58: 97–108.
EHRLICH, P. R. (1958): Problems of Arctic-Alpine insect distribution as illustrated by the butterfly genus Erebia (Satyridae). Proc. 10th Intern. Congr. Entomol.,: 1 683–686.
EHRLICH, P. R. EHRLICH, J. P. & HOLDREN, J. P. (1973): Human Ecology, Problems and Solutions. Freeman and Company, San Francisco.
EISENTRAUT, M. (1968): Die tiergeographische Bedeutung des Oku-Gebirges im Bamenda-Banso-Hochland (Westkamerun). *Bonn. Zool. Beitr.* 19: 170–175.
— (1970): Eiszeitklima und heutige Tierverbreitung im tropischen Westafrika. *Umschau* 3: 70–75.
— (1973): Die Wirbeltierfauna von Fernando Poo und Westkamerun. Bonner Zool. Monogr. 3.
EITSCHBERGER, U. & STEINIGER, H. (1973): Aufruf zur internationalen Zusammenarbeit an der Erforschung des Wanderphänomens bei den Insekten. *Atalanta* 4 (3): 133–192.
EKMAN, S. (1932): Prinzipielles über die Wanderungen und die tiergeographische Stellung des europäischen Aales, Anguilla anguilla (L.). *Zoogeographica* 1 (2): 85–106.
— (1935): Tiergeographie des Meeres, Leipzig.
— (1940): Biologische Geschichte der Nord- u. Ostsee. In: GRIMPE, Tierwelt der Nord- u. Ostsee, 1, Leipzig.
— (1940): Die schwedische Verbreitung der glazialmarinen Relikte. Verh. Int. Ver. Limnol. 9.
— (1953): Zoogeography of the Sea, London.
ELLENBERG, H. (1973): Die Ökosysteme der Erde. In: Ökosystemforschung. Verl. G. Fischer, Stuttgart.
ELSTER, H. J. (1968): Was ist 'Limnologie'? *Gas- und Wasserfach* 109: 651–652.
ELTON, C. S. (1966): The pattern of animal communities. Methuen, London. pp. 432.
EMERSON, A. E. (1952): The biogeography of termites. *Bull. Am. Mus. Nat. Hist.* 99: 217–225.
ENGLER, A. (1899): Die Entwicklung der Pflanzengeographie in den letzten 100 Jahren. Humboldt-Centenarschrift d. Ges. Erdkunde, Berlin, pp. 247.
ERGENZINGER, P. (1967): Rumpfflächen, Terrassen und Seeablagerungen im Süden des Tibesti-Gebirges. Abhdl. Dtsch. Geographentag.
— (1968): Vorläufiger Bericht über geomorphologische Untersuchungen im Süden des Tibesti-Gebirges. *Z. Geomorph.* 12 (1).
ERHART, H. (1956): Evolution des sciences. La genèse des sols entant que phenomène geologique. Masson & Cie., Paris.
ERICHSON, R. (1923): Die Mangrovevegetation. Halle.
ERNST, W. (1967): Bibliography on work on heavy-metal plant communities with the exception of serpentine. *Excerpt. bot. Sec.* 8: 50–61.
ESDORN, J. & PIRSON, H. (1973): Die Nutzpflanzen der Tropen und Subtropen in der Weltwirtschaft. Verl. G. Fischer, Stuttgart.
FABER, F. C. (1923): Zur Physiologie der Mangroven. *Ber. dtsch. bot. Ges.* 41.
FAIRBRIDGE, R. W. (1962): World Sea-Level and climatic changes. *Quaternaria* 6: 111–134.
FAIRHALL, A. W. (1973): Accumulation of Fossil CO_2 in the Atmosphere and the Sea. *Nature* 245: 20–23.

FILHA, I. G. (1971): Sôbre a Politica de Ocupação da Amazônia. A Amazonia Brasileira em Foco. Rio.
FITTKAU, E. J. & KLINGE, H. (1973): On Biomass and Trophic Structure of the Central Amazonian Rain Forest Ecosystem. *Biotropica* 5 (1): 2–14.
FITZSIMONS, V. F. M. (1962): Snakes of Southern Africa. Purnell and Sons LTD, Cape Town, Johannesburg.
FLINDT, R. & HEMMER, H. (1972): Paarungsrufe und das Verwandtschaftsproblem paläarktischer und nearktischer Anuren. *Biol. Zentralbl.* 91 (6): 699–706.
FLORA EUROPAEA (1972): University Press, Cambridge.
FOWLER, J. (1964): Gambusia fish for rice-field mosquito control in California. Proc. 17th Ann. Mtg. Utah Mosquito Abatement Assn.
FRANK, P. (1952): A laboratory study of intraspecies and interspecies competition in Daphnia pulicaria (FORBES) and Simocephalus vetulus (O. F. MÜLLER). *Physiol. Zool.* 25: 173–204.
— (1957): Coactions in laboratory populations of two species of Daphnia. *Ecology* 38: 510–519.
FRANZ, H. (1969): Vergleich der Hochgebirgsfaunen in verschiedenen Breiten der Westpaläarktis. *Verhdl. Dtsch. Zool. Ges.* Innsbruck.
— (1970): Die geographische Verbreitung der Insekten. *Hbd. Zool. Berlin* 4 (2): 1–111.
FREEMAN, T. N. (1972): A Correlation of some Butterfly Distributions with Geological Formations. *Can. Ent.* 104: 443–444.
FRITZSCHE, R., K., E., LEHMANN, W. & PROESELER, G. (1972): Tierische Vektoren pflanzenpathogener Viren. G. Fischer Verl. Stuttgart.
FUNKE, W. (1971): Food and Energy Turnover of Leafeating Insects and their Influence on Primary Production. *Ecol. Stud.* 2: 81–93.
— (1972): Energieumsatz von Tierpopulationen in Land-Ökosystemen. *Verhdl. Dtsch. Zool. Ges.* 65: 95–105.
GASKIN, D. E. (1972): Reappraisal of the New Zealand Mesozoic with Respect to Sea-Floor. Spreading and Modern Tectonic Plate Theory. XVII. Congrès internat. Zool. Biogéographie et liaisons inter-continentales au cours du Mésozoique. Monte Carlo (Manuskript).
GAUZE, G. F. (1934): Experimentelle Untersuchungen über den Kampf ums Dasein zwischen Paramaecium caudatum, Paramaecium aurelia und Stylonichia mytilus. *Zool. Z.* 13: 1–17.
— (1934): Eine mathematische Theorie des Kampfes ums Dasein und ihre Anwendung auf die Populationen von Hefezellen. *Bull. MOIP biol.* 43: 69–87.
— (1934): Über die Prozesse, durch die in den Infusorien Populationen eine Art durch eine andere ausgemerzt wird. *Zool. Z.* 13: 18–26.
— (1944): Einige Probleme der chemischen Biozönologie. *Usp. sovr. biol.* 17: 216–221.
GENTILLI, J. (1949): Foundations of Australian bird geography. *Emu* 49: 85–129.
GEORGE, W. (1962): Animal Geography. London, Melbourne.
GIGLIOLI, M. E. C. (1965): The influence of irregularities in the bush perimeter of the cleared agricultural belt around a gambian village on the flight range and direction of approach of a population of Anopheles gambiae melas. XII Int. Congr. Entomol. London, 757–758.
GLICK, P. A. (1939): The distribution of insects, spiders and mites in the air. *U. S. Dept. Agr. Techn. Bull.* 673.
GORMAN, G. C. & ATKINS, L. (1969): The Zoogeography of Lesser Antillean Anolis Lizards. *Bull. Mus. Comp. Zool.* 138 (3).
GOSLINE, W. A. (1972): A reexamination of the similarities between the freshwater fishes of Africa and South America. XVII. Congrès internat. Zool. Biogéographie et liaisons intercontinentales au cours du Mésozoique. Monte Carlo (Manuskirpt).
GRABERT, H. (1973): Die Biologie des Präkambrium. *Zbl. Geol. Paläont.* 5/6: 196–226.
GRAHAM, A. (1965): Origin and evolution of the biota of southeastern North America. Evidence from the fossil plant record. *Evolution* 18: 571–585.

— (1972): Floristics and Paleofloristics of Asia and Eastern North America. Elsevier Publ. Com., Amsterdam, London, New York.
GRAHAM, A. und JARZEN, D. (1969): Studies in Neotropical Palaeobotany. I. The Oligocene Communities of Puerto Rico. *Ann. Missouri Bot. Garden* 58: 308–357.
GREENWOOD, P. H. (1973): Morphology, endemism and speciation in African cichlid fishes. Verhdl. Dtsch. Zool. Ges. 66: 115–124.
GRESSITT, J. L. et al (1961): Problems in the Zoogeography of Pacific and Antarctic insects.
— (1965): Biogeography and Ecology of Land Arthropods of Antarctica. In: Biogeography and Ecology in Antarctica, 431–490, Junk, The Hague.
GREVE, P. A. & VERSCHUREN, H. G. (1971): Die Toxizität von Endosulfan für Fische in Oberflächengewässern. *Schrift. Ver. Wasser-, Boden-, Lufthygiene* 34.
GRISEBACH, A. (1866): Die Vegetationsgebiete der Erde. P.M. 11.
GROVE, A. T. und PULLAN, R. A. (1963): Some aspects of the Pleistocene Palaeogeographie of the Chad bassin. African Ecology and Human Evolution. *Viking Fund Publ. Anthropol.* 36: 230–245.
GULLAND, J. A. (1971): Ecological Aspects of Fishery Research. In: Advances in Ecological Research 7: 115–176. Acad. Press London and New York.
GÜNTHER, A. (1858): On the zoogeographical distribution of the reptiles. *Proc. Zool. Soc. London*: 21: 373–398.
GÜNTHER, K. (1969): Zur zoogeographischen Terminologie: Das 'Inselfaunenmuster' als zoogeographischer Begriff; die faunistischen 'Kleininselspezialisten'. *Wiss. Zeitschr. Ernst-Moritz-Arndt-Univers. Greifswald* 18: 9–15.
— (1970): Die Tierwelt Madagaskars und die zoogeographische Frage nach dem Gondwana-Land. *Sitz.-ber. Ges. naturf. Freunde, Berlin* (N.F.) 10: 79–92.
HAFFER, J. (1969): Speciation in Amazonian Forest Birds. *Science* 165: 131–137.
HAFFNER, W. (1971): Khumbu Himalaya. Landschaftsökologische Untersuchungen in den Hochtälern des Mt. Everest-Gebietes. *Erdwiss. Forsch.* 4: 244–263.
HALFTER, G. (1972): Elements anciens de l'entomofaune neotropicale: Ses implications biogeographiques. XVII Congrès internat. Zool. Biogéographie et liaisons intercontinentales au cours du Mésozoique. Monte Carlo (Manuskript).
HALL, B. P. & MOREAU, R. E. (1970): An atlas of speciation in African Passerine Birds. London.
HALLAM, A. (1973): Atlas of Palaeobiography. Elsevier Scient. Publ. Comp., Amsterdam, London, New York.
HARDY, R. N. (1972): Temperature and Animal Life. Camelot Press Ltd., London und Southampton.
HARRINGTON, H. J. (1965): Geology and Morphology of Antarctica. In: Biogeography and Ecology in Antarctica, 1–71, Junk, The Hague.
HAUDE, W. (1969): Erfordern die Hochstände des Toten Meeres die Annahme von Pluvial-Zeiten während des Pleistozäns? Meteor. Rdsch. 22 (2): 29–40.
HAVEN, ST. B. (1973): Competition for food between the intertidal Gastropods Acmaea scabra and Acmaea digitalis. *Ecology* 54 (1): 143–151.
HEATH, J. (1971): Instructions for Recorders. Biological Records Centre. Monks Wood Experimental Station Abbots Ripton. Huntingdonshire.
HEATH, J. und LECLERCQ, J. (1970): Erfassung der europäischen Wirbellosen. *Ent. Zeitschr.* 80 (19): 195–196.
HEILPRIN, A. (1887): The geographical and geological distribution of animals. Kegan Paul, Trench, London.
— (1907): The Geographical and Geological Distribution of Animals. London.
HENCKEL, P. A. (1963): On the ecology of the mangrove vegetation. *Mitt. flor.-soziol. Arbeitsgem.* 10.
HENNIG, W. (1957): Systematik und Phylogenese. Ber. Hundertjahrf. dtsch. ent. Ges., Berlin.

LYON, G. L. et al. (1970): Some trace elements in plants from serpentine soils. *New Zealand J. Sci.* 13: 133–139.
MACARTHUR, R. H. (1972): Geographical Ecology, Harper and Row, Publ., New York, Evanston, San Francisco, London.
MACARTHUR, R. H. & CONNEL, J. H. (1970): Biologie der Populationen. BLV, München, Basel, Wien.
MACARTHUR, R. H. & WILSON, E. O. (1963): An equilibrium theory of insular Zoogeography. *Evolution* 17: 373–387.
— (1971): Biogeographie der Inseln. Verl. Goldman, München.
MACDONALD, W. W. (1965): Mosquitoes and Disease in Sarawak. XII Int. Congr. Entomol. 820.
MAIN, A. R. LITTLEJOHN, M. J. & LEE. A. K. (1959): Ecology of Australian Frogs. In: Biogeography and Ecology in Australia. Verl. Junk, Den Haag.
MALEC, F. & STORCH, G. (1972): Der Wanderigel, Erinaceus algirus DUVERNOY & LEREBOULLET, 1842, und seine Beziehungen zum nordafrikanischen Herkunftsgebiet. *Säugetierkde. Mitt.* 20: 146–151.
MALOIY, G. M. O. (1972): Comparative Physiology of Desert Animals. Zool. Soc. London Acad. Press New York and London.
MARCUS, E, (1933): Tiergeographie. Potsdam.
MARTIN, P. S. & WRIGHT, H. E. (1967): Pleistocene Extinctions. Yale Univ. Press, New Haven, Conn.
MARTINI, F. (1965): Globale Verbreitung des Läuse-Rückfallfiebers. Welt-Seuchen-Atlas 2.
MATRONNE, E. DE, CHEVALLIER, A. & CUENOT, L. (1927): La distribution géographique des animaux. Paris.
MAYR, E. (1942): Systematics and the origin of species. Columbia Univ. Press, New York.
— (1943): The zoogeographic position of the Hawaiian Islands. *Condor* 45: 45–48.
— (1944): Wallace's line in the light of recent zoogeographic studies. *Quart Rev. Biol.* 19: 1–14.
— (1964): Inferences concerning the Tertiary American bird faunas. *Proc. Nat. Acad. Sci.* 51.
— (1963): The Fauna of North America, its origin and unique Composition. Proc. XVI Intern Congr. Zool., Washington, D.C.
— (1967): Artbegriff und Evolution. Paul Parey Verlag, Hamburg und Berlin.
MCBOYLE, G. R. (1972): Perception of urban climate. International Geography 1972; 22. *Int. Geogr. Congr.* 1: 162–164.
MCLUSKY, D. S. (1971): Ecology of estuaries. Heinemann Educ. Books Ltd. London.
MEISENHEIMER, J. (1904): Die bisherigen Forschungen über die Beziehungen der drei Südkontinente zu einem antarktischen Schöpfungszentrum. *Naturw. Wsch.* 3.
— (1915): Zoogeographie. In: Handwörterbuch d. Naturwiss. 10, Jena.
MELL, R. (1929): Leitgedanken zu einer Ökologie ostasiatischer Reptilien, insbesondere Schlangen. X. Congrès Int. Zool. Budapest, 1470–1477.
— (1958): Zur Geschichte der ostasiatischen Lepidopteren. *Deutsche Ent. Z.* 5: 185–213.
MERRIAM, C. . (1892): The geographic distribution of life in North America with special reference to Mammalia. *Proc. Biol. Soc. Wash.* 7: 1–64.
— (1894): Laws of temperature control of geographic distribution of terrestiral mammals and plants. *Nat. Geogr. Mag.* 6: 229–238.
— (1898): Life zones and crop zones in the United States. *U.S. Dept. Agr. Bull.* 10 pp. 79.
MESSERLI, B. (1967): Die eiszeitliche und die gegenwartige Vergletscherung im Mittelmeerraum. Geogr. Helvetica, 105–228.
MESSERLI, B. (1972): Formen und Formungsprozesse in der Hochgebirgsregion des Tibesti. Hochgebirgsforschung 2: 23–86.
MERTENS, R. (1952): Die Amphibien und Reptilien von El Salvador. Abhdl. Senck. Naturf. Ges. 487.
— (1948): Die Tierwelt des tropischen Regenwaldes. Verl. W. Kramer, Frankfurt.

— (1961): Die tiergeographischen Beziehungen Australiens zu anderen Festländern. *Geogr. Rdsch.* 13 (3): 99–105.
— (1961): Tier und Landschaft. *Frankf. Geogr. H.* 37: 31–85.
— (1972): Madagaskars Herpetofauna und die Kontinentaldrift. *Zool. Mededel.* 46 (7): 91–98.
MEYER, H. (1974): Wälder des Ostalpenraumes. Verl. G. Fischer, Stuttgart.
MEYER, P. (1972): Zur Biologie und Ökologie des Atlashirsches Cervus elaphus barbarus, 1833. *Z. Säugetierkd.* 37: 101–116.
MICHAELSEN, J. W. (1903): Die geographische Verbreitung der Oligochaeten. Friedländer, Berlin. pp. vi + 186.
MÖBIUS, K. (1877): Die Auster und die Austernwirtschaft. Berlin.
MÖBIUS, K. (1891): Die Tiergebiete der Erde, ihre kartographische Abgrenzung und museologische Bezeichnung. *Arch. Naturgesch.* 57: 277–291.
MOHR, E. und DUNKER, G. (1930): Vom 'Formenkreis' Mus musculus. *Zool. Jb. Syst.* 56: 65–72.
MONOD, T. (1963): The late Tertiary and Pleistocene in the Sahara and adjacent Southerly Regions. African Ecology and Human Evolution. Viking F. Publicat. *Anthropology* 36: 117–230.
— (1972): Sur la distribution de quelques Crustaces Malacostraces d'eau douce ou saumatre. XVII Congrès internat. Zool. Biogéographie et liaisons inter-continentales au cours du Mésozoique. Monte Carlo (Manuskript).
MOORE, D. M. (1972): Connections between Cool Temperate Floras, with particular Reference to Southern South America. In: Taxonomy, Phytogeography and Evolution. Acad. Press, London, New York.
MOREAU, R. E. (1931): Pleistocene Climatic Changes and the Distribution of Life in East Africa. *J. Ecol.* 21: 415–435.
— (1963): Vicissitudes of the African biomes in the Late Pleistocene. *Proc. Zool. Soc. London.* 141: 395–421.
— (1966): The Bird Faunas of Africa and its islands. Academic Press, London and New York.
— (1969): Climatic changes and the distribution of forest vertebrates in West Africa. *J. Zool.* 158: 39–61.
— (1972): The Palaearctic – African Bird Migration Systems. Acad. Press, London und New York.
MOURSI, A. (1962): The letal doses of CO_2, N, NH3 and H_2S for soil arthropods. *Pedobiologia* 2: 9–14.
MÜLLER, G. (1966): Die Sedimentbildung im Bodensee. Naturwissenschaften 53. Berlin.
MÜLLER, P. (1968): Die Herpetofauna der Insel von São Sebastião (Brasilien). Verl. Saarbr. Zeit., Saarbrücken.
— (1969): Herpetologische Beobachtungen auf der Insel Marajó. *DATZ* 22 (4): 117–121.
— (1969): Zur Verbreitung von Hemidactylus mabouia (MOREAU DE JONÉS) auf den südbrasilianischen Inseln. *Zool. Anz.* 182 (3/4): 196–203.
— (1969): Zur Verbreitung der Gattung Chironius (Serpentes/Colubridae) auf den südbrasilianischen Inseln. *Senck. biol.* 50 (3/4): 133–141.
— (1969): Einige Bemerkungen zur Verbreitung von Vipera aspis (Serpentes/Viperidae) in Spanien. *Salamandra* 5 (1/2): 57–62.
— (1970): Vertebratenfaunen brasilianischer Inseln als Indikatoren für glaziale und postglaziale Vegetationsfluktuationen. *Abhdl. Dtsch. Zool. Ges. Würzburg* 1969: 97–107.
— (1970): Durch den Menschen bedingte Arealveränderungen brasilianischer Wirbeltiere. *Nat. u. Museum* 100 (1): 22–37.
— (1971): Ausbreitungszentren und Evolution in der Neotropis. *Mitt. Biogeogr. Abt. Geogr. Inst. Univers. Saarl.* 1: 1–20.
— (1971): Biogeographische Probleme des Saar-Mosel-Raumes, dargestellt am Hammelsberg bei Perl. *Faun.-flor. Not. aus Saarl.* 4 (1/2): 1–14.
— (1971): Herpetologische Reiseeindrücke aus Brasilien. Salamandra.

— (1972): Die Bedeutung der Ausbreitungszentren für die Evolution neotropischer Vertebraten. Zool. Anzeiger.
— (1972): Centres of Dispersal and Evolution in the Neotropical Region. *Stud. Neotr. Fauna* 7: 173–185.
—(1972): Der neotropische Artenreichtum als biogeographisches Problem. Festbundel Brongersma. *Zool. Med.* 47: 88–110.
— (1972): Biogeography and Evolution in South America. Int. Geogr. Congr. Montreal.
— (1972): Die Bedeutung der Biogeographie für die ökologische Landschaftsforschung. *Biogeographica* 1: 25–53.
— (1972): Die Bedeutung biogeographischer Methoden für die Bearbeitung saarländischer Umweltprobleme. Umwelt-Saar 1972: 28–40.
— (1972): Biogeographie und die 'Erfassung der Europäischen Wirbellosen'. *Ent. Zeitschr.*. 82 (3).
— (1973): Probleme des Ökosystems einer Industriestadt, dargestellt am Beispiel von Saarbrücken. Belastung und Belastbarkeit von Ökosystemen. Giessen.
— (1973): Erziehung zum Umweltbewußtsein in der Universität. Umwelt-Saar 1973: 68–78.
— (1973): Die Verbreitung der Tiere. Grzimeks Tierleben, 16. Kindler Verl.
— (1973): Amazonische Nationalparks. *Entomol. Zeitschr.* 83 (6): 57–64.
— (1973): The Dispersal Centres of Terrestrial Vertebrates in the Neotropical realm. *Biogeographica* 2: 1–243. Junk, The Hague.
— (1973): Historisch-biogeographische Probleme des Artenreichtums der südamerikanischen Regenwälder. Amazoniana.
— (1974): Beiträge der Biogeographie zur Geomedizin und Ökologie des Menschen. Verl. Steiner, Wiesbaden.
— (1973): Die Erfassung der europäischen Fauna als europäische Aufgabe. *Mitt. Biogeogr. Abt. Univers. Saarlandes* 5: 1–2.
— (1973): Monomorphismus und Polymorphismus italienischer Chalcides chalcides – Populationen. *Salamandra* 9 (1): 13–17.
— (1974 ed.): Abhandlungen des Gesellschaft für Ökologie. Junk, The Hague.
MÜLLER, P. und SCHMITHÜSEN, J. (1970): Probleme der Genese südamerikanischer Biota. Festschr. Hirt, Kiel.
MUNROE, E. (1965): Zoogeography of Insects and allied Groups. *Ann. Rev. Entomol.* 10.
MURRAY, A. (1866): The geographical distribution of mammals. London.
MYERS, G. S. (1965): Gambusia, the fish destroyer. Tropical Fish Hobbyist.
NADIG, A. (1968): Über die Bedeutung der Massifs de Refuge am südlichen Alpenrand (dargelegt am Beispiel einiger Orthopterenarten). *Mitt. schweiz. ent. Ges.* 41: 341–358.
NASH, T. A. M. (1969): Africas Bane. The Tsetse Fly. London.
NAUMANN, C. (1969): Untersuchungen zur Systematik und Phylogenese der holarktischen Sesiiden (Insecta, Lepidoptera). Inaugural-Dissertation, Bonn.
NEILL, W. T. (1969): The geography of life. New York.
NELSON, G. J. (1969): The problem of Historical Biogeography. *Syst. Zool.* 18 (2): 243–246.
NEUMANN, D. (1974): Zielsetzungen der Physiologischen Ökologie. Verkl. Ges. Ökel., Junk, The Hague.
NEVO, E. et al. (1972): Competitive Exclusion between Insular Lacerta Species (Sauria, Lacertidae). Notes on Experimental Introductions. *Oecologia* 10: 183–190.
NEWBIGIN, M. I. (1936): Plant and animal geography. Methuen, London, pp. 298.
— (1968): Plant and animal geography, London and New York.
NIETHAMMER, G. (1958): Tiergeographie. In: Fortschritte der Zoologie. 11, G. Fischer Verl., Stuttgart.
— (1959): Die Rolle der Auslese bei Wüstenvögeln. *Bonn. Zool. Beitr.* 10: 179–197.
— (1969): Die wertvollsten Vögel der Welt. *Vogel-Kosmos* 9: 304–309.
NIETHAMMER, G. and KRAMER, H. (1966): Tiergeographie. In: Fortschritte der Zoologie 18. G. Fischer Verl., Stuttgart.

NIETHAMMER, J. (1972): Der Igel von Teneriffa. *Zool. Beitr.* 18 (2): 307–309.
NIKLFELD, H. (1972): Bericht über die Kartierung der Flora Mitteleuropas. *Taxon* 20 (4): 545–571.
NUORTEVA, P. (1963): Die Rolle der Fliegen in der Epidemiologie der Poliomyelitis. *Anzeiger für Schädlingskd.* 36: 149–155.
ØKLAND, F. (1955): Generell Dyregeografi. Aschehoug, Oslo. pp. 166.
— (1956): Tiergeographie – Ökologie. *Biol. Zentralbl.* 75: 1–2, 83–85.
ORTMANN, A. E. (1896): Grundzüge der marinen Tiergeographie. Jena.
— (1901): The theories of the origin of the Antarctic faunas and floras. *Amer. Nat.* 35.
OSBORN, H. F. (1910): The age of mammals in Europe, Asia and North America. New York.
OSCHE, G. (1958): Beiträge zur Morphologie, Ökologie und Phylogenie der Ascarioidea, Nemetoda. Z. f. Parasitenkd. 18: 479–572.
OVERBECK, J. (1972): Zur Struktur und Funktion des aquatischen Ökosystems. Ber. dtsch. Botan. Ges. 85: 533–577.
PACKARD, A. S. (1880): Summary of locust flights from 1877–1879. *2nd Rep. U.S. Entom. Comm. Ch.* 7: 160–163.
PACKARD, A. S. & THOMAS, C. C. (1878): Migrations. *1st Rep. U.S. Entom. Comm. Ch.* 7: 143–211.
PAIJMANS, K. und LÖFFLER, E. (1972): Hight-Altitude Forests and Grasslands of Mt. Albert Edwards, New Guinea. *J. Trop. Geogr.* 35: 58–64.
PALMEN, E. (1944): Die Anemohydrochore Ausbreitung der Insekten als zoogeographischer Faktor. *Ann. Zool. Soc. Zool. Bot. Fenn. Vanamo* 10 (1): 1–262.
PALMER, A. R. (1973): Cambrian Trilobites. In: Atlas of Palaeobiogeography. Elsevier Scient. Publ. Comp. Amsterdam, London, New York.
PATTERSON, C. (1972): The distribution of Mesozoic freshwater fishes. XVII Congrès internat. Zool. Biogéographie et liaisons inter-continentales au cours du Mésozoique. Monte Carlo. (Manuskript).
PAULIAN, R. (1972): La position de Madagascar dans le double problème du peuplement animal et des translations continentales. XVII Congrès internat. Zool. Biogéographie et liaisons inter-continentales au cours du Mésozoique. Monte Carlo (Manuskript).
PEAKE, J. F. (1971): The evolution of terrestrial faunas in the western Indian Ocean. *Phil. Trans. Roy. Soc. Lond.* 260: 581–610.
PEARY, J. und CASTENHOLZ, R. W. (1964): Temperature strains of a thermophilic blue-green alga. *Nature, Lond.* 202: 720–721.
PERRING, F. H. (1963): Data-processing for the Atlas of the British Flora. *Taxon* 12: 183–190.
— (1965): Mapping the flora of Europe. *Bot. Tidskr.* 61: 328–332.
PETERS, G. (1972): Chorologische und phylogenetische Aspekte in der Variabilität des Flügelgeäders einiger Arten der Sympetrum-Gruppe. *Dtsch. Ent. Z.* 19: 263–286.
PETERS, H. (1949): Fliegen- und Rattenbekämpfung – wichtige Aufgaben der Stadthygiene. *G.I.* 9/10: 160–169.
PIANKA, E. R. (1972): Zoogeography and Speciation of Australian Desert Lizards: An Ecological Perspective. *Copeia* 1972 (1): 127–145.
PIETSCH, M. (1970): Vergleichende Untersuchungen an Schädeln nordamerikanischer und europäischer Bisamratten (Ondatra zibethicus L. 1766). *Z. Säugetierkd.* 35 (5): 257–288.
PIMENOV, M. G. (1968): The analysis of the distribution of species of Angelica occurring in the Soviet Far East. *Bot. Zurn. Moscow, Leningrad,* 53 (7): 932–946.
PITELKA, F. A. (1951): Speciation and ecologic distribution in American jays of the genus Aphelocoma. *Univ. Calif. Publ. Zool.* 50: 195–464.
PLUMSTEAD, E. P. (1973): The Late Palaeozoic Glossopteris Flora. In: Atlas of Palaeobiogeography. Elsevier Scient. Publ. Comp. Amsterdam, London, New York.
POLL, M. & LELEUP, N. (1965): Un Poisson aveugle nouveau de la famille des Brotulidae provenant des Iles Galapagos. *Bull. Cl. SC. Acad. roy. Belg.* 51 (4): 464–474.

POLL, M. & VAN MOL, J. J. (1966): Au Sujet d'une espèce inconnue de Brotulidae littoral des îles Galapagos, apparentée à l'espèce aveugle Caecogilbia galapagosensis POLL et LELEUP. *Bull. Cl. Sc. Acad. roy. Belg.* 52: 1444–1461.
PRAKASH, I. (1974): The ecology of vertebrates of the Indian Desert. In Ecology and Biogeography in India. Junk, The Hague.
PRENANT, M. (1933): Géographie des animaux. Colin, Paris.
PRESTON, F. W. (1962): The canonical distribution of commoness and rarity. *Ecology* 43: 185–215, 410–432.
PRETZMANN, G. (1965): Quantitative field-studies on the cycle of Tick-Borne Encephalitis in a focus in Lower Austria. XII Int. Congr. Entomol. London, 776.
PREVOST, J. & SAPIN-JALOUSTRE, J. (1965): Ecologie des Manchots antarctiques. In: Biogeography and Ecology in Antarctica, 551–648, Junk, The Hague.
PROBALD, F. (1972): Deviations in the heat balance: the basis of Budapest's urban climate. Internat. Geography 1972; 22. *Int. Geogr. Congr.* 1: 184–186.
PRYDE, PH. R. (1972): Conservation in the Soviet Union. Cambridge Univ. Press.
PRYOR, L. D. (1959): Species Distribution and Association in Eucalyptus. In: Biogeography and Ecology in Australia. Junk, The Hague.
PUTHZ, V. (1972): Revision of the Stenus – Species of New Guinea. Part. II (Coleoptera: Staphylinidae). *Pacific Insects* 14 (3): 475–527.
RATCLIFFE, F. N. (1959): The Rabbit in Australia. In: Biogeography and Ecology in Australia. Verl. Junk, Den Haag.
RAUSER, J. (1970): Biogeographische Landschaftsforschung und ihre Bedeutung für die geographische Praxis. Quaestiones geobiologica .
REED, A. CH. (1970): Extinction of Mammalian Megafauna in the Old World late Quaternary. *Bio-Science* 20 (5): 284–288.
REICHENOW, A. (1888): Die Begrenzung zoogeographischer Regionen vom ornithologischen Standpunkt aus. *Zool. Jahrb. (Syst.)* 3: 661–704.
REID, J. A. (1965): The species in medical Entomology, with examples from Mosquitoes. XII Int. Congr. Entomol. London, 759–760.
REINIG, W. F. (1937): Die Holarktis. Jena.
— (1939): Real Systematic Units in Zoology and their genetic structure. *Research and Progress* 5 (1): 20–39.
REISE, D. (1972): Untersuchungen zur Populationsdynamik einiger Kleinsäuger unter besonderer Berücksichtigung der Feldmaus, Microtus arvalis (Pallas, 1779). *Z. Säugetierkd.* 37: 65–97.
REMANE, A. (1943): Die Bedeutung der Lebensformtypen für die Ökologie. Biologia Generalis 17: 164–182.
REMMERT, H. (1972): Die Tundra Spitzbergens als terrestrisches Ökosystem. *Umschau* 72 (2): 41–44.
RENSCH, B. (1931): Tiergeographie (Literatur von 1908–1930). Geograph. Jahrb. 45: 51–132.
— (1950): Verteilung der Tierwelt im Raum. In: BERTALANFFY, Hdb. Biol. 5, Potsdam.
REYMENT, R. (1972): 'Biogéographie et liaisons internat. au cours du Mésozoique' XVII. Congrès internat. Zool. Biogéographie et liaisons inter-continentales au cours du Mésozoique. Monte Carlo (Manuskript).
RIDLEY, H. N. (1930): The Dispersal of Plants throughout the World. Reeve & Co., LTD., Ashford, Kent.
RODE, W. W. (1913): Schutzeinrichtungen von Früchten und Samen gegen die Einwirkung fliessenden Meerwassers. Inaugural-Dissertation. Gleiwitz.
RODENWALDT, (1925): Malaria und Küstenform. *Arch. Schiffs- und Tropenhyg.* 29: 292–304.
RUBIN, M. J. (1965): Antarctic climatology. In: Biogeography and Ecology in Antarctica 72–96, Junk, The Hague.
RUETIMEYER, L. (1867): Über die Herkunft unserer Tierwelt. Eine zoogeographische Skizze. Georg, Basel.

RUSTAMOV, A. K. (1955): Beiträge zum Begriff 'Lebensformen' in der Ökologie der Tiere. *Zool. Z.* 34: 710–718.
SALOMONSEN, F. (1930): Diluviale Isolation und Artenbildung. Proc. VII. Int. Ornith. Congress Amsterdam.
— (1955): The evolutionary significance of Bird-Migration. *Det Kongelige Danske Videnskabernes Selskab* 22 (6): 1–62.
SAUER, E. G. F. und ROTHE, P. (1972): Ratite Eggshells from Lanzarote, Canary Islands. *Science* 176: 43–45.
SAVAGE, J. M. (1966): The origins and history of the Central American Herpetofauna. *Copeia* 1966 (4): 719–766.
SCHÄFER, A. (1974): Die Bedeutung der Saarbelastung für die Arealdynamik von Molluskenpopulationen. Abhdl. Ges. f. Ökologie. Junk, The Hague.
SCHERF, H. (1969): Zahl, Vorkommen und Verbreitung der Läuse. *Nat. Rdsch.* 22: 399–400.
SCHILDER, F. A. (1956): Lehrbuch der allgemeinen Zoogeographie. Jena.
SCHMARDA, L. K. (1853): Die geographische Verbreitung der Tiere. Gerond, Vienna. pp, vii + 755.
SCHMID, R. (1969): Weltweite Grippe-Epidemie. *Naturw. Rdsch.* 22 (2): 114–115.
SCHMIDT, G. (1969): Vegetationsgeographie auf ökologisch-soziologischer Grundlage. Teubner Verlagsges., Leipzig.
SCHMIDT, K. P. (1954): Faunal realms, regions, and provinces. *Quart. Rev. Biol.* 29: 322–31.
SCHMINKE, H. K. (1972): Mesozoic international relationships as evidenced by Bathynellid Crustacea. XVII. Congrès internat. Zool. Biogéographie et liaisons intercontinentales au cours du Mésozoique. Monte Carlo (Manuskript).
SCHMITHÜSEN, J. (1957): Anfänge und Ziele der Vegetationsgeographie. *P.M.* 2: 81–92.
— (1966): Problems of Vegetation history in Chile and New Zealand. *Vegetatio* 18.
— (1968): Allgemeine Vegetationsgeographie. 3. Aufl. Walter de Gruyter & Co., Verl., Berlin.
SCHNEIDER, B. (1971): Das Tyrrhenisproblem. Interpretation auf zoogeographischer Grundlage. Dargestellt an Amphibien und Reptilien. Inaugural-Dissertation, Saarbrücken.
SCHOLANDER, P. F. (1968): How mangrove desalinate seawater? Physiol. Plant. 21: 251–261.
SCHOLZ, H. (1967): Baumbestand, Vegetations-Gliederung und Klima des Tibesti-Gebirges. *Berl. Geogr. Abhdl.* 5: 11–16.
SCHÖNBECK, H. (1969): Eine Methode zur Erfassung der biologischen Wirkung von Luftverunreinigungen durch transplantierte Flechten. Staub-Reinhalt. *Luft* 29: 14–18.
SCHÜZ, E. (1971): Grundriß der Vogelzugkunde. Verl. Paul Parey, Hamburg, Berlin.
SCHWABE, G. H. (1970): Surtsey, Island. Natürliche Erstbesiedlung (Ökogenese) der Vulkaninsel. Schrift. Naturwiss. Ver. Schleswig-Holstein, 1–120.
SCHWABE, H. & BEHRE, K. (1971): Ökogenese der Insel Surtsey 1968 bis 1970. *Naturwiss. Rdsch.* 24 (12): 513–519.
SCHWOERBEL, J. (1971): Einführung in die Limnologie. G. Fischer Verl., Stuttgart.
SCHWEINFURTH, U. (1966): Neuseeland. Beobachtungen und Studien zur Pflanzengeographie und Ökologie der antipodischen Inselgruppe. Bonner Geogr. Abhdl., Bonn.
SCHWERDTFEGER, F. (1968): Ökologie der Tiere. 3. Demökologie. Verl. P. Parey, Hamburg u. Berlin.
SCHREIBER, H. (1973): Ausbreitungszentren von Sphingiden (Lepidoptera) in der Neotropis. Amazonia 4(3): 273–281.
SCLATER, P. L. (1858): On the general geographical distribution of the members of the class Aves. *J. Proc. Linn. Soc. London (Zool.)* 2: 130–145.
— (1874): The Geographical distribution of mammals, Manchester.
SEGERSTRALE, S. G. (1954): The freshwater amphipods, Gammarus pulex (L.) and Gammarus lacustris (G.O. Sars) in Danmark and Fennoskandia – a contribution to the late – and postglacial immigration history of the aquatic fauna of Northern Europe. Soc. Sci. Fenn. Comment. Biol. 15.

— (1957): On the immigration of the glacial relicts of Northern Europe, with remarks on their prehistory. *Soc. Sci. Fenn. Comment. Biol.* 16.
— (1962): The immigration and prehistory of the glacial relicts of Eurasia and North America. A Survey and discuss. of modern views. *Int. Rev. Hydrobiol.* 47.
— (1966): Adaptational problems involved in the history of the glacial relicts of Eurasia and North America. *Rev. Roum. Biol.* 11 (1): 59–67.
SELANDER, R. K. (1971): Systematics and speciation in Biology. Advanced Biology. Acad. Press, New York, London.
SHELFORD, V. E. (1911): Physiological animal geography. *J. Morphol.* 22: 551–618.
SHOTTON, F. W. (1967): The problems and contributions of methods of absolute dating within the Pleistocene period. *Quart. J. Geol. Soc.* 122: 357–383.
SHUGART, H. H. & PATTEN, B. C. (1972): Niche Quantification and the Concept of Niche Pattern. In: Systems Analysis and Simulation in Ecology. Acad. Press. New York und London.
SIMBERLOFF, D. S. (1969): Experimental zoogeography of Islands: a model for insular colonization. *Ecology* 50: 296–314.
SIMBERLOFF, D. S. & WILSON, E. O. (1969): Experimental Zoogeography of Islands: the colonization of empty Islands. *Ecology* 50: 278–296.
SIMPSON, G. G. (1940): Review of the mammal bearing Tertiary of South America. *Proc. Amer. Philos. Soc.* 83 (5): 649–709.
— (1950): History of the fauna of Latin America. *Amer. Scient.* 38: 361–389.
— (1953): Evolution and geography. Eugene Ore.
— (1965): The geography of Evolution. Philadelphia, New York.
— (1966): Mammalian evolution on the southern continents. *Neues Jb. Geol. Paläontol.* 125: 1–18.
SIOLI, H. (1954): Gewässerchemie und Vorgänge in den Böden im Amazonasgebiet. *Naturwissenschaften* 41 (19): 456–457.
— (1955): Beiträge zur regionalen Limnologie des brasilianischen Amazonasgebietes III. Über einige Gewässer des oberen Rio Negro-Gebietes. *Arch. Hydrobiol.* 50 (1): 1–32.
— (1968): Zur Ökologie des Amazonasgebietes. Biogeography and Ecology in South America. 1: 137–169. Junk, The Hague.
SMARDA, J. (1970): Studium der Verbreitung von Pflanzenarten als Unterlage zur Klassifikation der Landschaftstypen. *Quaestiones geobiologicae* 7: 123–127.
SMETANA, A. (1965): Einfluss der Umweltfaktoren auf die Wirtsspezifität der Kleinsäugerläuse (Anoplura). XII. Internat. Congr. Entomol. London, 828–829.
SOPER, J. H. (1966): Machine-plotting of phytogeographical data. *Canad. Geogr.* 10: 15–26.
SOUTHWICK, CH. H. (1972): Ecology and the Quality of our Environment. Van Nostrand Reinhold Company. New York, Cincinnati, Toronto, London, Melbourne.
SPATZ, H. & STEPHAN, H. (1961): Adaptive Konvergenz von Schädel und Gehirn bei 'Kopfwühlern'. *Zool. Anz.* 166 (9/12): 402–423.
SPERLICH, D. (1973): Populationsgenetik. G. Fischer Verl., Stuttgart.
STEFFAN, A. W. (1972): Zur Produktionsökologie von Gletscherbächen in Alaska und Lappland. Verhandl. *Dtsch. Zool. Gesellsch.* 65: 73–78.
STERBA, G. (1962): Die Neunaugen (Petromyzonidae). Hdb. Binnenfischerei Mitteleurop. 3. Stuttgart.
STEUBING, L. (1970): Arbeitsber. region. Planungsgem. Untermain 34–37, Frankfurt.
STOCKER, O. (1962): Steppe, Wüste und Savanne. Festschr. Firbas. *Veröff. Geobot. Inst. Rübel* 37.
STODDART, D. R. (1972): Geography and the ecological approach: the Ecosystem as a geographic principle and method. In: The Conceptual Revolution in Geography. Univ. London Press Ltd.
STOLL, O. (1901): Über Xerothermische Relikte in der Schweizer Fauna der Wirbellosen. Festschr. Geogr. Ethnogr. Ges. Zürich.

STOP-BOWITZ, C. (1969): Did lumbricids survive the quaternary glaciations in Norway? *Pedobiologia* 9: 93–98, Jena.
STROHL, G. (1921): Physiologische Gesichtspunkte in der Tiergeographie. *Vierteljahrschr. Naturf. Gesellsch. Zürich.* 66: 1–22.
STUGREN, B. (1972): Grundlagen der allgemeinen Ökologie. G. Fischer Verl., Jena.
SUESS, E. (1909): Das Antlitz der Erde. Wien und Leipzig.
SUKOPP, H. (1972): Wandel von Flora und Vegetation in Mitteleuropa unter dem Einfluß des Menschen. Umweltschutz in Land- und Forstwirtschaft 50: 112–139, Verl. P. Parey, Hamburg u. Berlin.
SUKACEV, V. N. (1960): Gegenseitiges Verhältnis der Begriffe Biogeozönose, Ökosystem und Facies. Pocvovedenie 6: 1–10.
SWAINSON, W. (1835): A treatise on the geography and classification of animals. Longman, Green, London. pp. 367.
TANAI, T. (1972): Tertiary History of Vegetation in Japan. In: Floristics and Paleofloristics of Asia and Eastern North America. Elsevier Publ. Com., Amsterdam, London, New York.
TANSLEY, A. (1935): The use and abuse of vegetational concepts and terms. *Ecology* 16: 284–307.
TARLING, D. H. & RUNCORN, S. K. (1973): Implications of continental Drift to the earth sciences. Academic Press, London and New York.
TENOW, O. (1972): The outbreaks of Oporinia autumnata Bkh. and Operophthera spp. (Lep., Geometridae) in the Scandinavian mountain chain and northern Finland 1862–1968. Zool. Bidr. fran Uppsala 2.
THENIUS, E. (1972): Säugetierausbreitung in der Vorzeit. Geophysik ermöglicht neue Einsichten. *Umschau* 72 (5): 148–153.
THIELCKE, G. (1965): Gesangsgeographische Variation des Gartenbaumläufers (Certhia brachydactyla) im Hinblick auf das Artbildungsproblem. *Z. Tierpsychologie* 22 (5): 542–566.
— (1969): Geographic variation in bird vocalizations. Bird Vocalizations, Cambridge Univ. Press.
THIELE, H. U. (1961): Zuchtversuche an Carabiden, ein Beitrag zu ihrer Ökologie. *Zool. Anz.* 167: 9–12.
— (1964): Experimentelle Untersuchungen über die Ursachen der Biotopbindung bei Carabiden. *Z. Morphol. Ökol. Tiere* 53: 387–452.
— (1964): Ökologische Untersuchungen an bodenbewohnenden Coleopteren einer Heckenlandschaft. *Z. Morphol. Ökol. Tiere* 53: 537–586.
— (1967): Ein Beitrag zur experimentellen Analyse von Euryökie und Stenökie bei Carabiden. *Z. Morphol. Ökol. Tiere* 58: 355–372.
— (1968): Was bindet Laufkäfer an ihre Lebensräume. *Naturwiss. Rundsch.* 21 (2): 57–65.
THIELE, H. U. & KOLBE, W. (1962): Beziehungen zwischen bodenbewohnenden Käfern und Pflanzengesellschaften in Wäldern. *Pedobiologia* 1: 157–173.
THIELE, H. U. & LEHMANN, H. (1967): Analyse und Synthese im tierökologischen Experiment. *Z. Morph. Ökol. Tiere* 58: 373–380.
THIENEMANN, A. (1914): Die Entstehung neuer Tierformen durch die Eiszeit. *Die Naturwissenschaften* 2: 581–587.
— (1928): Die Reliktenkrebse Mysis relicta, Pontoporeia affinis, Pallasea quadrispinosa und die von ihnen bewohnten norddeutschen Seen. *Arch. f. Hydrobiol.* 19: 521–582.
— (1950): Verbreitungsgeschichte der Süßwassertierwelt Europas, Binnengewässer 18, Stuttgart.
THOMAS, W. A. (1972): Indicators of environmental quality. Plenum Press, New York and London.
THOMAS, W. L. (1956): Man's Role in Changing the Face of the Earth. Univ. Chicago Press.
TISCHLER, W. (1955): Synökologie der Landtiere, Stuttgart.
TOBIAS, W. (1972): Ist der Schlammröhrenwurm Branchiura sowerbyi BEDDARD 1892 (Oli-

gochaeta: Tubificidae) ein tropischer Einwanderer im Untermain? *Nat. u. Mus.* 102 (3): 93–107.
TRALAU, H. (1973): Some Quaternary Plants. In: Atlas of Palaeobiogeography. Elsevier Scient. Publ. Comp. Amsterdam, London, New York.
TROLL, C. (1936): Termiten-Savannen. Länderkdl. Forsch. Festschr. N. Krebs. Stuttgart.
— (1962): Die dreidimensionale Landschaftsgliederung der Erde. Hermann von Wissmann-Festschr., Tübingen.
TROUESSART, E. L. (1890): La géographie zoologique. Bailliere, Paris. pp.
TURNER, D. B. (1964): A diffusion model for an urban area. *J. Appl. Met.* 3.
UCKO, P. J. & DIMBLEBY, G. W. (1969): The domestication and exploitation of plants and animals. Gerald Dockworth & Co., London.
UDVARDY, M. D. F. (1964): Bird faunas of North America. Proc. 13th Int. Ornith. Congress 1147–1167.
— (1968): The concept of faunal dynamism and the analysis of an example. *Bonner zool. Beitr.* 20: 1–10.
— (1969): Dynamic Zoogeography with special reference to land animals. Van Nostrand Reinh. Comp. New York.
URQUHARDT, F. A. (1960): The monarch butterfly. Univ. Toronto Press, Toronto.
VALENTINE, D. H. (1972): Taxonomy, Phytogeography and Evolution. Acad. Press, London und New York..
VANDEL, A. (1972): Les isopodes terrestres et la Gondwana. XVII. Congrès internat. Zool. Biogéographie et liaisons inter-continentales au cours du Mésozoique. Monte Carlo (Manuskript).
UVAROV, B. P. (1921): A revision of the genus Locusta with a new theory as to the periodicity and migration of locusts. *Bull. Ent. Res.* 12: 135–163.
— (1951): Locust Research and Control (1929–1950). Colonial Research Publication No. 10, London.
VAN MOL, J. J. (1967): Ecologie comparée de deux espèces de Brotulidae (Pisces) des îles Galapagos: Caecogilbia deroyi POLL et VAN MOL 1967 et C. galapagosensis Poll et Leleup 1965. *Bull. Cl. Acad. roy. Belg.* 53 (3): 232–248.
VAN STEENIS, C. G. G. J. (1972): Nothofagus, Key Genus to Plant Geography. In: Taxonomy, Phytogeography and Evolution. Acad. Press, London, New York.
VANZOLINI, P. E. & WILLIAMS, E. E. (1970): South American Anoles: The Geographic differentiation and evolution of the Anolis chrysolepsis species (Sauria, Iguanidae). Arq. Zool. 19: 1–298.
VARGA, Z. (1970): Extension, Isolation, Micro-Evolution. *Acta Biologica Debrecina* 7/8: 193–209.
— (1974): Geographische Isolation und Mikroevolution bei den Lepidopteren der Hochgebirge der Balkanhalbinsel. Habilitationsarbeit, Debrecen (in press).
VAUFREY, R. (1929): Les éléphants nains des îles méditerran. et la question des isthmes pleistocènes. Masson et Cie. Paris.
VAURIE, C. (1951): A study of Asiatic larks, *Bull. Amer. Mus. Nat. Hist.* 97: 431–526.
VERNADSKIJ, V. J. (1967): Die Biosphäre. Moskau.
VIELLIARD, J. (1972): Définition du Bécasseau variable Calidris alpina (L.). *Alauda* 40 (4): 321–342.
VOOUS, K. H. (1965): Antarctic birds. In: Biogeography and Ecology in Antarctica, 649–689, Junk, The Hague.
VOS, A. DE, MANVILLE, R. H. and GELDER, R. G. VAN (1956): Introduced Mammals and their influence on native Biota. *Zoologica* 41: 163–194, New York.
VUILLEUMIER, F. (1970): Insular biogeography in continental regions. 1. The northern Andes of South America. *Amer. Natur.* 104: 373–388.
— (1970): Speciation in South American birds: 1 progress report. *Act. IV Congr. Latin. Zool.* 1: 239–255.

WACE, N. M. (1965): Vascular plants. In: Biogeography and Ecology in Antarctica, 201–266, Junk, The Hague.

WAGNER, A. (1844): Die geographische Verbreitung der Säugetiere. *Abhandl. Bayer. Akad. Wiss., Math. Phys. Cl. IV* (1): 1–146, (2): 1–108, (3): 3–114.

WAGNER, G. (1967): Beiträge zum Sauerstoff-, Stickstoff- und Phosphathaushalt des Bodensees. *A. für Hydrobiol.* 63.

WAGNER, M. (1868): Die Darwin'sche Theorie und das Migrationsgesetz der Organismen. Dunker und Humblot, Leipzig.

— (1889): Die Entstehung der Arten durch räumliche Sonderung. Basel.

WAIBEL, L. (1912): Physiologische Tiergeographie. *Hettners G. Zeitschr.* 18.

— (1921): Urwald, Veld, Wüste. Breslau.

WALKER, D. (1972): Bridge and Barrier: The Natural and Cultural History of Torres Strait. Australian National Univ. Press, Canberra.

WALLACE, A. R. (1860): On the zoological geography of the Malay Archipelago. *Proc. Lin. Soc. Zool. London* 4: 173–184.

— (1876): Geographical distribution of animals, London.

WALTER, H. (1971): Biosphäre, Produktion der Pflanzendecke und Stoffkreislauf in ökologisch-geographischer Sicht. *G.Z.* 59 (2): 116–130.

— (1973): Allgemeine Geobotanik. Ulmer Verl., Stuttgart.

WANGERIN, W. (1912): Reliktenbegriff und Konstanz der Pflanzenstandorte. Festschr. Preuss. Bot. Verein.

WARNECKE, G. (1927): Gibt es xerothermische Relikte unter den Makrolepidopteren des Oberrheingebietes von Basel bis Mainz? *Arch. Insektenkd. Oberrheingeb. und angrenz. Ld.* 2 (3): 81–119.

— (1936): Über die Konstanz der ökologischen Valenz einer Tierart als Voraussetzung für zoogeographische Untersuchungen. *Entomol. Rundschau* 53: 203–206, 217–219, 230–232.

— (1950): Wanderfalter in Europa. *Z. Wien Ent. Ges.* 35: 100–106.

— (1961): Rezente Arealvergrößerung bei Makrolepidopteren in Mittel- und Nordeuropa. *Bonn. Zool. Beitr.* 12: 113–141.

WEBER, M. (1902): Der Indo-australische Archipel und die Geschichte seiner Tierwelt. Verl. G. Fischer, Jena.

WEGENER, A. (1929): Die Entstehung der Kontinente und Ozeane. Sammlung Vieweg, Braunschweig.

WHITLEY, G. P. (1959): The Freshwater Fishes of Australia. In: Biogeography and Ecology in Australia. Junk, The Hague.

WIELGOLASKI, F. E. & ROSSWALL, TH. (1972): Tundra Biome. Proceeding of IV Intern-Meeting on the Biological Productivity of Tundra, Leningrad, USSR, 1971. Stockholm.

WIJMSTRA, T. A. (1969): Palynology of the first 30 meters of a 120 m deep section in northern Greece. *Acta Botan. Neerl.* 18: 511–527.

WILLIAMS, L. G. (1972): Plankton diatom species biomasses and the Quality of American Rivers and the Great Lakes. *Ecology* 53 (6): 1038–1050.

WILLIS, J. C. (1922): Age and area. Cambridge Univ. Press, Cambridge. pp. 259.

WILSON, E. O. & BOSSERT, W. H. (1973): Einführung in die Populationsbiologie. Springer, Berlin, Heidelberg, New York.

WILSON, E. O. & SIMBERLOFF, D. S. (1969): Experimental zoogeography of Islands: defaunation and monitoring techniques. *Ecology* 50: 267–278.

WITTMANN, O. (1934): Die biogeographischen Beziehungen der Südkontinente. Die antarktischen Beziehungen. *Zoogeographica* 2 (2): 246–304.

WOLFE, J. A. & HOPKINS, D. M. (1967): Climatic changes recorded by Tertiary land floras in northwestern North America. In: Tertiary correlations and climatic changes in the Pacific. Sasaki, Sendai.

WOLFENBARGER, D. O. (1946): Dispersion of small organisms: distance dispersion rates of

bacteria, spores, seeds, pollen and insects; incidence rates of diseases and injuries. *Amer. Midland Naturalist* 35: 1–152.
WOLTERECK, R. (1928): Über die Spezifität des Lebensraumes, der Nahrung und der Körperformen bei pelagischen Cladoceren und über 'Ökologische Gestaltsysteme'. Biol. Zentralbl. 28: 521–551.
WRIGHT, H. E. & FREY, D. G. (1965): The Quaternary of the United States. Princeton Univ. Press, Princeton.
YOUNGNER, V. B. & MCELL, C. M. (1972): The Biology and Utilization of Grasses. Acad. Press, New York und London.
YURTSEV, B. A. (1972): Phytogeography of Northeastern Asia and the Problem of Transberingian Floristic Interrelations. In: Floristics and Paleofloristics of Asia and Eastern North America. Elsevier Publ. Comm., Amsterdam, London, New York.
ZABELIN, J. M. (1959): Theorie der physischen Geographie. Moskau.
ZIEGLER, H. (1969): Physiologische Anpassungen der Pflanzen an extreme Umweltbedingungen. *Naturwiss. Rdsch.* 22 (6): 241–247.
ZIMMERMAN, E. C. (1948): Insects of Hawaii. Univ. Hawaii Press, Honolulu.

SUBJECT INDEX